T0212778

SpringerBriefs in Electrical and Computer Engineering

More information about this series at http://www.springer.com/series/10059

Nan Cheng • Xuemin (Sherman) Shen

Opportunistic Spectrum Utilization in Vehicular Communication Networks

 Springer

Nan Cheng
Department of Electrical
  and Computer Engineering
University of Waterloo
Waterloo, ON, Canada

Xuemin (Sherman) Shen
Department of Electrical
  and Computer Engineering
University of Waterloo
Waterloo, ON, Canada

ISSN 2191-8112   ISSN 2191-8120   (electronic)
SpringerBriefs in Electrical and Computer Engineering
ISBN 978-3-319-20444-4   ISBN 978-3-319-20445-1   (eBook)
DOI 10.1007/978-3-319-20445-1

Library of Congress Control Number: 2015942246

Springer Cham Heidelberg New York Dordrecht London

Printed on acid-free paper

Springer International Publishing AG Switzerland is part of Springer Science+Business Media (www.springer.com)

Preface

VehiculAr NETworks (VANETs) have been envisioned to improve road safety and efficiency, and provide Internet access on the move, by incorporating wireless communication and informatics technologies into the road transportation system. VANETs can facilitate a myriad of attractive applications related to road safety (e.g., collision avoidance, safety warnings, and remote vehicle diagnostic), infotainment (e.g., web browsing, file downloading, and video streaming), and traffic efficiency improvement and traffic management (e.g., real-time traffic notification and electronic tickets). Through the evolving VANETs applications and services, not only the road safety and efficiency can be greatly enhanced, but also the driving and in-vehicle experiences can be improved.

In this monograph, we focus on the utilization of opportunistic spectrum bands, including both unlicensed bands and licensed bands for VANETs, in order to improve the performance of VANETs and facilitate more data-craving applications. The research is of great importance since VANETs are facing spectrum scarcity problem due to the dramatic growth of mobile data traffic and the limited bandwidth of dedicated vehicular communication band. In Chap. 1, we overview VANETs and describe the problem of spectrum scarcity. In Chap. 2, we provide a comprehensive survey on existing opportunistic spectra utilization methods. In Chap. 3, we study the utilization of licensed spectrum through cognitive radio. Specifically, the probability distribution of the channel availability is first derived by means of a finite-state continuous-time Markov chain (CTMC), jointly considering the mobility of vehicles, and the spatial distribution and the temporal channel usage pattern of primary transmitters. Using the channel availability statistics, we propose a game theoretic spectrum access scheme for vehicles to opportunistically access licensed channels in a distributed manner. In Chap. 4, we investigate on the data delivery through WiFi access network in vehicular communication environment, focusing on the analysis of delay and throughput performance. In specific, we consider a generic vehicular user having Poisson data service arrival to download/upload data from/to the Internet via sparsely deployed WiFi networks (want-to) or the cellular network providing full service coverage (have-to). In this scenario, the WiFi offloading performance, characterized by *offloading effectiveness*, is analyzed under the requirement of a desired *average service delay* which is the time the data services can be deferred for WiFi availability. We establish the theoretical relation between offloading effectiveness and average service delay by an M/G/1/K queueing model, and the tradeoff is examined. Finally, conclusions and future research directions are given in Chap. 5. This monograph validates the feasibility of using opportunistic spectra (cognitive radio and WiFi) for VANETs, and also evaluates the performance of such opportunistic vehicular communication paradigms. Therefore, this monograph can provide valuable insights on the design and deployment of future VANETs.

We would like to thank Prof. Jon W. Mark, Dr. Ning Lu, Dr. Ning Zhang, Dr. Tom H. Luan, from Broadband Communications Research (BBCR) Group at the University of Waterloo, and Prof. Tingting Yang from Navigation College at Dalian Maritime University, for their contributions in the presented research works. We also would like to thank all the members of BBCR group for the valuable discussions and their insightful suggestions, ideas, and comments. Special thanks are also due to the staff at Springer Science+Business Media: Jennifer Malat and Melissa Fearon, for their help throughout the publication preparation process.

Waterloo, ON, Canada   Nan Cheng
                                                       Xuemin (Sherman) Shen

Contents

Acronyms

AP	Access Point
BS	Base Station
CR	Cognitive Radio
CRN	Cognitive Radio Network
DSRC	Dedicated Short Range Communications
D2D	Device-to-device
EST	Effective Service Time
GPS	Global Positioning System
ITS	Intelligent Transportation System
LTE	Long Term Evolution
MAC	Medium Access Control
MNO	Mobile Network Operator
NE	Nash Equilibrium
PDF	Probability Density Function
PT	Primary Transmitter
PU	Primary User
QoS	Quality of Service
RSU	RoadSide Unit
SU	Secondary User
VANETs	VehiculAr NETworks
V2I	Vehicle-to-infrastructure
V2V	Vehicle-to-vehicle
VU	Vehicular User

Chapter 1
Introduction

Vehicular networks play a critical role in both developing the intelligent transportation system and providing data services to vehicular users (VUs) by incorporating wireless communication and informatics technologies into the transportation system. However, due to the dramatic growth of mobile data traffic and the limited bandwidth of dedicated vehicular communication band, vehicular networks are facing spectrum scarcity problem in which spectrum resource is not sufficient to satisfy the data requirements, and thus the performance is compromised. In this chapter, we first overview the vehicular networks, and then describe the spectrum scarcity problem in vehicular networks, including the causes, and the impacts of the problem on the performance of vehicular networks. At last, the aim of the monograph is provided.

1.1 Overview of Vehicular Networks

As an indispensable part of modern life, motor vehicles have continued to evolve since people expect more than just vehicle quality and reliability. With the rapid development of information and communication technologies, equipping automobiles with wireless communication capabilities is the frontier in the evolution to the next generation intelligent transportation systems (ITS). In the last decade, the emerging VehiculAr NETworks (VANETs) have attracted much interest from both academia and industry, and significant progress has been made. VANETs are envisioned to improve road safety and efficiency and provide Internet access on the move, by incorporating wireless communication and informatics technologies into the transportation system. VANETs can facilitate a myriad of attractive applications, which are usually divided into two main categories: safety applications (e.g., collision avoidance, safety warnings, and remote vehicle diagnostic [1, 2]) and

© The Author(s) 2016
N. Cheng, X. (Sherman) Shen, *Opportunistic Spectrum Utilization in Vehicular Communication Networks*, SpringerBriefs in Electrical and Computer Engineering, DOI 10.1007/978-3-319-20445-1_1

infotainment applications (e.g., file downloading, web browsing, and audio/video streaming [3, 4]). To support these various applications, the U.S. Federal Communication Commission (FCC) has allocated totally 75 MHz in the 5.9 GHz band for Dedicated Short Range Communications (DSRC), based on the legacy of IEEE 802.11 standards (WiFi). On the other hand, the car manufacturers, suppliers and research institutes in Europe have initialed the Car-to-Car Communication Consortium (C2C-CC) with the main objective of utilizing inter-vehicle communication to increase road safety and efficiency. IEEE has also developed IEEE 1609 family, which consists of standards for wireless access in vehicular environments (WAVE).

Unlike most mobile ad hoc networks studied in the literature, VANETs present unique characteristics, which impose distinguished challenges on networking. (a) *Potential large scale*: VANETs are extremely large-scale mobile networks, which can extend over the entire road network with a great amount of vehicles and roadside units; (b) *High mobility*: the movement of vehicles make the environment in which the VANET operators extremely dynamic. On highways, vehicle speed of over 150 km/h may occur, while in the city, the speed may exceed 60 km/h while the node density may be very high, especially during rush hour; (c) *Partitioned network*: the high mobility of vehicles may lead to large inter-vehicle distance in sparse scenarios, and thus the network is usually partitioned, consisting of isolated clusters of nodes; (d) *Network topology and connectivity*: the scenario of VANETs is very dynamic because vehicles are constantly moving and changing their position. Therefore, the network topology changes very often and the links between vehicles connect and disconnect frequently. In addition, the links are also affected by the unstable outdoor wireless channels; (e) *Varied applications*: applications of VANETs are of a large variety and with different quality of service (QoS) requirements. All these features dramatically complicate network protocol design, implementation and performance evaluation.

VANETs basically consist of two types of communications, i.e., vehicle-to-vehicle (V2V) communications and vehicle-to-infrastructure (V2I) communications [5], as shown in Fig. 1.1. Installed with on-board units (OBUs), vehicles can communicate with each other in ad hoc manner without the assistance of any built infrastructure, which is referred to as V2V communications. By disseminating information such as location, speed, and emergency warning messages to nearby vehicles using V2V communications, VANETs can support varied applications such as public safety applications, vehicular traffic coordination, road traffic management [6], and some comfort applications (e.g., interactive gaming, and file sharing) [7, 8], etc. In February 2014, the U.S. Department of Transportation announced that it would begin to take steps to enable V2V communication technology for light vehicles by early 2017. Communications between vehicles and communication infrastructure (usually offers Internet access) are referred to as V2I communications. Internet access has become an essential part of people's daily life, and thus is required anywhere and anytime. It is evidenced that the demand for high-speed mobile Internet services has increased dramatically. A recent survey reveals that Internet access is predicted to become a standard feature of future motor vehicles [9]. Providing high-rate Internet access for vehicles can not only meet

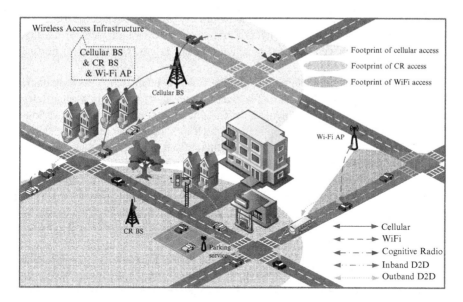

Fig. 1.1   An overview of vehicular networks and available communication spectra

the ever-increasing Internet data demand of travelers, such as multi-media services, but also enrich some safety-related applications, such as intelligent anti-theft and tracking [10], online vehicle diagnosis [11], and so forth. Besides, jointly using both V2V and V2I communications has attracted much attention since it can provide better performance [12, 13].

Motivated by the vision and prospect of VANETs, both the academia, industry and government institutions have done numerous activities. A review of past and ongoing related programs and projects in USA, Japan and Europe can be found in [1]. The standards of VANETs is reviewed in [6]. There are also a lot of research works on VANETs, which have been surveyed in papers such as [1, 14].

1.2   Spectrum Scarcity in VANETs

The FCC has allocated 75 MHz spectrum to DSRC, and wireless wide area network (WWAN) can be a practical and seamless way to provide Internet connectivity to vehicles [15], such as off-the-shelf 3G and Long Term Evolution (LTE) cellular networks. However, VANETs still face the problem of spectrum scarcity, which has been demonstrated in [16]. The primary reasons of spectrum scarcity might be: (1) the ever-increasing data intensive applications, such as high-quality video streaming and user generated content (UGC), require a large amount of spectrum resources, and thereby the quality of service (QoS) is difficult to satisfy merely by the dedicated bandwidth; (2) the number of connected vehicles and devices is soaring, and thus the requirement for communication bandwidth increases dramatically. In urban environments, the spectrum scarcity is more severe due to high vehicle density, especially in some places where the vehicle density is much higher than normal [8, 17].

Growth of demands: People tend to require richer contents when they are static as well as on the road. The types of services required by people in the car have turned from simple GPS, navigation, in-car phone and email to more various services, featuring multimedia applications such as video/audio streaming, UGC upload and sharing, online gaming, web surfing, etc. It is predicted that over two-third of the global mobile data traffic will be video by 2018. These multimedia applications often require large communication bandwidth, for example, the size of a typical high definition movie is 5.93 GB while an Android game may need 1.8 GB download/upload to play [18]. In addition, it is reported that the average speed of mobile connection will surpass 2 Mbps by 2016, and the smartphones will generate 2.7 GB of data traffic on average per month.

Connected vehicles and devices: There are two types of entities in VANETs that generate and consume data. The first type is connected vehicles that integrated with Internet access capability and services. It is predicted that the percentage of Internet-integrated vehicle services will jump from 10 % today to 90 % by 2020 [19]. The connected vehicles can offer a number of integrated services to drivers (e.g., real-time navigation, driver assistance, online diagnosis, etc.) as well as to passengers (e.g., e-mails, video on demand, etc.). The other type of entities are the mobile terminals of in-vehicle passengers. It is reported that the connected mobile devices have become more than the world's population by the end of 2013. Mobile users expect to be connected anywhere and anytime, even when they are traveling in vehicles.

Cellular communication technologies can provide reliable and ubiquitous Internet access services and deliver data traffic for VANETs. Although 4G cellular technologies such as LTE-A have extremely efficient physical and MAC layer protocols, the cellular network nowadays is straining to meet the current mobile data demand [20]; on the other hand, the explosive growth of mobile data traffic is no end in sight, resulting in an increasingly severe overload problem. Consequently, simply using cellular infrastructure for vehicle Internet access may worsen the overload problem, and degrade the service performance of both non-vehicular and VUs. For DSRC, comparing with the large mobile data demand, the bandwidth of DSRC is limited. In urban environments, the spectrum scarcity is more severe due to high vehicle density, especially in some places where the vehicle density is much higher than normal [27, 28]. Moreover, due to the contention-based channel access model, the performance of vehicular mobile data services cannot be guaranteed as in the cellular-based technologies. In summary, the dedicated DSRC spectrum and the cellular network may not be sufficient to provide a huge number of VUs with high-quality services, and thus other solutions are required.

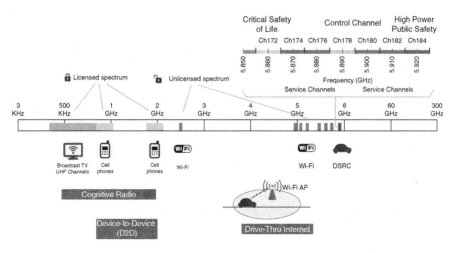

Fig. 1.2 Opportunistic spectrum bands for VANETs

1.3 Aim of the Monograph

As mentioned above, the dedicated 5.9 GHz band is not enough to satisfy the data requirements of vehicular networks, while the cellular network is already congested and may be very expensive to use. Other than DSRC and the cellular network, there are several opportunistic spectrums that can be utilized for VANETs. The three main opportunistic spectrums that can be utilized for VANETs are: (1) licensed spectrum (e.g., TV white band) that can be utilized through cognitive radio technology, (2) ISM spectrum that can be utilized by WiFi, and (3) cellular spectrum that can be opportunistically utilized through device-to-device communications, as shown in Fig. 1.1. The spectrum bands are shown in Fig. 1.2.

Cognitive radio (CR) is a possible complementary technology which allows users to communicate opportunistically on spatially and/or temporally vacant licensed radio spectrum for other communication systems. The IEEE 802.11af [21] and the IEEE 802.22 [22] standards take advantage of dynamic spectrum access (DSA) on TV white space to support wireless local area networks (WLANs) and wireless regional area networks (WRANs), respectively.

With millions of hotspots deployed all over the world, **WiFi**, operating on unlicensed spectrum is a complementary solution to deliver data content at low cost. The feasibility of WiFi for outdoor Internet access at vehicular mobility has been demonstrated in [23], referred to as drive-thru Internet. Recent advances in Passpoint/Hotspot 2.0 powered by WiFi Alliance make WiFi more competitive to provide secure connectivity and support seamless roaming. Different from the cellular network, WiFi cannot provide fully coverage based on the deployment of APs/hotspots, and thus the spectrum is spatially opportunistic for vehicles to use.

As a promising solution to offload the cellular network (CN), **device-to-device (D2D)** communication technology has gained much attention recently [24].

The basic tenet of D2D communications is that mobile users in proximity can communicate directly with each other on the cellular spectrum (or other spectrum bands) without traversing the base station or the backhaul networks. By utilizing the proximity of mobile users and direct data transmission, D2D communications can increase spectral efficiency and throughput, and reduce communication delay for mobile users [25], which may be applied to many VANETs applications such as video streaming, location-aware advertisement, safety related applications and so forth. However, the spectrum for D2D communication is opportunistic in the way that D2D communication should avoid interfering the uplink/downlink cellular communication and the D2D communications of neighboring devices since they may use the same spectrum resources. For example, it is shown in [26] that the probability of having D2D links increases with the pathloss component because the larger pathloss component implies weaker interference caused by D2D transmissions to the cellular base station. The D2D communication is not allowed when the required transmit power may cause interference to the cellular uplink/downlink transmission higher than the minimal interference threshold.

CR, WiFi, and D2D communication have received extensive research attentions, and have been proved to be capable of supporting broadband communication for static or mobile users. However, the research works on utilizing such opportunistic spectrum for VANETs are limited, considering the unique features of VANETs aforementioned. A better understanding of how VANETs can effectively utilize the spectrum opportunities will shed light not only to the design and implementation of related protocols and mechanism, but also to economics issues, such as where to deploy business WiFi hotspots and how to decide operator's price strategy, etc., which motivates our work.

The aim of this monograph is to investigate how to utilize opportunistic spectra for VANETs considering different scenarios and applications. Specifically, we try to address the following research issues: (a) The features and characteristics of spectrum opportunities for VANETs; and (b) how much data can be delivered by exploiting the opportunistic spectrum. To answer these questions, in this monograph, we analyze the spectrum availability jointly considering the characteristics of the spectrum and the mobility of vehicles, and investigate the throughput and delay performance of VANETs using the opportunistic spectra. Based on the investigations on these issues, we can elaborate the insights and implications for design and deployment of future VANETs.

References

1. Karagiannis G, Altintas O, Ekici E, Heijenk G, Jarupan B, Lin K, Weil T (2011) Vehicular networking: a survey and tutorial on requirements, architectures, challenges, standards and solutions. IEEE Commun Surv Tutorials 99:1–33
2. Omar H, Zhuang W, Li L (2013) VeMAC: a TDMA-based MAC protocol for reliable broadcast in VANETs. IEEE Trans Mob Comput 12(9):1724–1736

3. Luan T, Cai L, Chen J, Shen X, Bai F (2014) Engineering a distributed infrastructure for large-scale cost-effective content dissemination over urban vehicular networks. IEEE Trans Veh Technol 63(3):1419–1435

4. Trullols-Cruces O, Fiore M, Barcelo-Ordinas J (2012) Cooperative download in vehicular environments. IEEE Trans Mob Comput 11(4):663–678

5. Kenney J (2011) Dedicated short-range communications (DSRC) standards in the United States. Proc IEEE 99(7):1162–1182

6. Moustafa H, Zhang Y (2009) Vehicular networks: techniques, standards, and applications. Auerbach Publications, Boston

7. Bai F, Krishnamachari B (2010) Exploiting the wisdom of the crowd: localized, distributed information-centric VANETs. IEEE Commun Mag 48(5):138–146

8. Lu R, Lin X, Luan T, Liang X, Shen X (2012) Pseudonym changing at social spots: an effective strategy for location privacy in VANETs. IEEE Trans Veh Technol 61(1):86–96

9. KPMG's global automotive executive survey (2012) [Online]. Available: http://www.kpmg.com/GE/en/IssuesAndInsights/ArticlesPublications/Documents/Global-automotive-executive-survey-2012.pdf

10. Ramadan M, Al-Khedher M, Al-Kheder S (2012) Intelligent anti-theft and tracking system for automobiles. Int J Mach Learn Comput 2(1):88–92

11. Lin J, Chen S, Shih Y, Chen S (2009) A study on remote on-line diagnostic system for vehicles by integrating the technology of OBD, GPS, and 3G. World Acad Sci Eng Technol 56:56

12. Cheng X, Yang L, Shen X, D2D for intelligent transportation systems: a feasibility study. IEEE Trans Intell Transp Syst (to appear)

13. Zheng K, Liu F, Zheng Q, Xiang W, Wang W (2013) A graph-based cooperative scheduling scheme for vehicular networks. IEEE Trans Veh Technol 62(4):1450–1458

14. Hartenstein H, Laberteaux K (2008) A tutorial survey on vehicular ad hoc networks. IEEE Commun Mag 46(6):164–171

15. Chen B, Chan M (2009) Mobtorrent: a framework for mobile internet access from vehicles. In: Proceedings of IEEE INFOCOM, Rio de Janeiro, April 2009

16. Ghandour AJ, Fawaz K, Artail H (2011) Data delivery guarantees in congested vehicular ad hoc networks using cognitive networks. In: Proceedings of IEEE IWCMC, pp 871–876

17. Lu N, Luan T, Wang M, Shen X, Bai F (2012) Capacity and delay analysis for social-proximity urban vehicular networks. In: Proceedings of IEEE INFOCOM, Orlando, March 2012

18. The 1000x mobile data challenge (2013) [Online]. Available: http://www.qualcomm.com/media/documents/files/1000x-mobile-data-challenge.pdf

19. Connected Car Industry Report (2014) [Online]. Available: http://blog.digital.telefonica.com/connected-car-report-2014/

20. Asadi A, Wang Q, Mancuso V (2014) A survey on device-to-device communication in cellular networks. IEEE Commun Surv Tutorials 16(4):1801–1819

21. Flores AB, Guerra RE, Knightly EW, Ecclesine P, Pandey S (2013) IEEE 802.11 af: a standard for TV white space spectrum sharing. IEEE Commun Mag 51(10):92–100

22. Stevenson CR, Chouinard G, Lei Z, Hu W, Shellhammer SJ, Caldwell W (2009) IEEE 802.22: the first cognitive radio wireless regional area network standard. IEEE Commun Mag 47(1):130–138

23. Bychkovsky V, Hull B, Miu A, Balakrishnan H, Madden S (2006) A measurement study of vehicular internet access using in situ Wi-Fi networks. In: Proceedings of ACM MobiCom, USA, September 2006

24. Doppler K, Rinne M, Wijting C, Ribeiro C, Hugl K (2009) Device-to-device communication as an underlay to lte-advanced networks. IEEE Commun Mag 47(12):42–49

25. Golrezaei N, Molisch AF, Dimakis AG (2012) Base-station assisted Device-to-Device communications for high-throughput wireless video networks. In: Proceedings of IEEE ICC, Ottawa, June 2012

26. Johnson DB, Maltz DA (1996) Dynamic source routing in ad hoc wireless networks. In: Kluwer international series in engineering and computer science. Springer, New York, pp 153–179
27. Cheng N, Zhang N, Lu N, Shen X, Mark J, Liu F (2014) Opportunistic Spectrum Access for CR-VANETs: A Game-Theoretic Approach. IEEE Trans Veh Technol 63(1):237–251
28. Lu R, Lin X, Luan T, Liang X, Shen X (2012) Pseudonym changing at social spots: An effective strategy for location privacy in VANETs. IEEE Trans Veh Technol 61(1):86–96

Chapter 2
Opportunistic Communication Spectra Utilization

The objective of the monograph is to utilize the opportunistic spectra for VANETs through technologies such as CR, WiFi, and D2D communications. This chapter introduces the background and surveys the literature of these technologies, focusing on research works related to VANETs.

2.1 Cognitive Radio (Licensed Bands) for VANETs

Cognitive radio is a promising approach to deal with the spectrum scarcity, which enables unlicensed users to opportunistically exploit the spectrum owned by licensed users [1, 2]. In cognitive radio networks (CRNs), licensed users and unlicensed users are typically referred to as primary users (PUs) and secondary users (SUs), respectively. Specifically, SUs perform spectrum sensing before transmission, through which they can identify and exploit spectrum opportunities without interfering with the transmissions of PUs. By means of CR, not only can CRNs provide better QoS for SUs, but also the spectrum utilization is significantly improved. The main research topics of CRNs are spectrum sensing, spectrum sharing, and spectrum decision (or spectrum access), which have been extensively studied. The literature surveys for general CR networks and technologies can be found in [3–5]. As an example of the application of CR technology, the under-utilized TV white spaces (TVWS), which include VHF/UHF frequencies, have been approved for non-TV communications in many countries such as USA and Canada, using an emerging technology named Super WiFi. Several standards have been established for Super WiFi, such as IEEE 802.22 and IEEE 802.11af [6].

A natural question then rises whether CR can be applied to solve the problem of spectrum scarcity for VANETs. Recent researches in the literature demonstrate its feasibility [7–10]. With CR technology, VANETs have been coined as CR-VANETs,

© The Author(s) 2016
N. Cheng, X. (Sherman) Shen, *Opportunistic Spectrum Utilization in Vehicular Communication Networks*, SpringerBriefs in Electrical and Computer Engineering, DOI 10.1007/978-3-319-20445-1_2

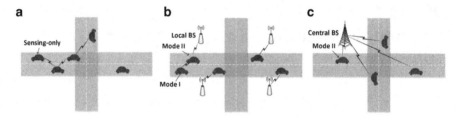

Fig. 2.1 Three deployment architecture of CR-VANETS with different device operations

whereby vehicles can opportunistically access licensed spectrum owned by other systems, outside the IEEE 802.11p specified standard 5.9-GHz band, such as digital television (DTV) and cellular networks. Considering the highly dynamic mobility, vehicles are expected to exploit more spatial and temporal spectrum opportunities along the road than stationary SUs. Other than simply placing a CR in vehicles, CR-VANETs has many unique features that should be considered. Different from static CR networks in which the spectrum availability is only affected by the spectrum usage patterns of the primary network, in CR-VANETs, the spectrum availability perceived by vehicles is also a function of the mobility of vehicles. Therefore, spectrum sensing should be conducted over the movement path of vehicles, leading to a spatiotemporal distribution, rather than temporal only. In addition, the constrained nature of vehicle mobility according to street patterns can be utilized. For example, the spectrum information in other locations can be obtained by a vehicle through information exchange with vehicles moving from those locations. And a vehicle can adapt its operations in advance using such information and its predicted movement.

FCC defined various device operations of CR devices in [11], motivated by spectrum database access capability, mobility and awareness of location. Different operation modes are associated with different deployment architecture of CR-VANETs, which are shown in Fig. 2.1.

- Spectrum database: Database which stores the usage information of TVWS. User can query the spectrum database to check what frequencies can be used, for a given location, without causing harmful interference to primary systems.
- Sensing-only mode: Devices cannot access spectrum database, and can access the spectrum relying only on the spectrum sensing result, such as vehicles in Fig. 2.1a. Cooperation can be used to improve the accuracy of spectrum sensing.
- Mode I: Devices with no geolocation and access capability to spectrum database. However, they can query Mode II devices for spectrum information updates, such as Mode I vehicles in Fig. 2.1b.
- Mode II: Devices are aware of location (e.g., via global positioning system (GPS) device) and capable of accessing spectrum database, such as Mode II vehicles in Fig. 2.1b, c.

2.1.1 Spectrum Sensing in CR-VANETs

2.1.1.1 Per-Vehicle Sensing

In per-vehicle sensing, vehicles sense the channels using the traditional sensing techniques, i.e., matched filter detection, energy detection, and cyclostationary feature detection [3]. Per-vehicle has the advantage that the implementation complexity and network support is minimal since each vehicle senses the spectrum and makes decision individually. However, the accuracy could be low given the high mobility of vehicles and the obstructed environments that may cause shadowing and fading effects. In [12], a mechanism is proposed to improve the accuracy of spectrum sensing by exploring the signal correlation between TV and 2G cellular channels. They prove that when the signals from adjacent TV and cellular transmitters are received in a common place, a strong Received Signal Strength Indicator (RSSI) can be detected. As a result, by comparing the signal with the fluctuations of cellular channels, a sudden change in the TV band can be verified.

2.1.1.2 Geolocation-Based Sensing

As discussed above, FCC has recommended to use the location information and spectrum database for CR users. The spectrum database can provide information about the bands, such as the types, locations and specific protection requirements of PUs, the availability of the bands, etc. Assisted by the spectrum database, vehicles can adjust the transmission parameters to avoid interfering PUs without sensing. Since most vehicles are equipped with localization systems (e.g., GPS), geolocation-based sensing is suitable for vehicles. Some spectrum databases are already available for users to access and query, for example, the TV query service in the United States [13] and Google spectrum database (shown in Fig. 2.2). In [14], spectrum database assisted CR-VANETs are proposed, in which fixed BSs are deployed along the road and provide spectrum database access to nearby vehicles. The deployment density of BSs is optimized to minimize the average cost of VUs accessing spectrum database while guarantee a low level of error of estimating available spectrum. The simulation results indicate that the cost of accessing spectrum information increases with the density of BSs since more vehicles are querying the spectrum database via BSs, which is a more expensive and accuracy way than obtaining the spectrum database information from nearby vehicles or spectrum sensing. In [15], a geo-location database approach is used to create spectrum availability map on I-90 highway in the state of MA. Then, a discussion on the number of non-contiguous blocks, the number of available channels, and design of transceivers is followed.

However, several concerns about the spectrum database may include the cost of building and maintaining the database, the coverage area of the service, significant query overhead, etc. In [12], it is proposed to jointly use the spectrum database and

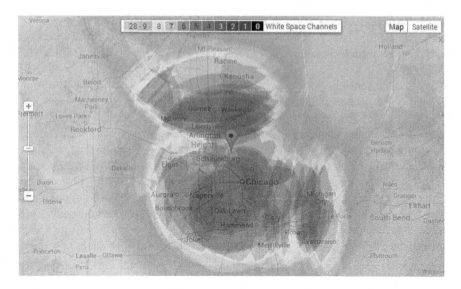

Fig. 2.2 TV white bands usage around Chicago from Google spectrum database. *Colored areas* correspond to channels that are used (Color figure online) https://www.google.com/get/spectrumdatabase/channel/

spectrum sensing. The signal correlation between TV and 2G cellular channels and the mobility of vehicles are utilized to reduce the number of queries to spectrum database in order to save the cost in terms of both money and communication overhead. A vehicle can send the channel measurements to nearby vehicles, containing the degree of correlation between TV and cellular channels. When other vehicles arrive the same location, they can conduct spectrum sensing, and make a decision on whether spectrum information update is necessary based on measurements sent by former vehicles in this location. The results show that about 23 % queries are reduced, which can benefit spectrum database deployment.

2.1.1.3 Infrastructure-Based Sensing

A sensing coordination framework which utilizes road side units (RSUs) is proposed in [16]. Unlike centralized sensing schemes in which a centralized controller gathers sensing reports from all users and allocates the channels to users, in [16], multiple RSUs are responsible to assist and coordinate the spectrum sensing and access for nearby vehicles. RSUs continuously detect the occupancy of PUs at its location, and then the coarse detection results are sent to vehicles. Vehicles conduct the fine-grained sensing and access the channel correspondingly. The results indicate that the sensing coordination framework outperforms the stand-alone sensing scheme in terms of sensing overhead, successful sensing rate, probability of sensing conflict, etc. One advantage of such schemes is that the change of government policies and

PU parameters can be easily loaded in RSUs, which can further adapt the operation and save the cost.

2.1.1.4 Cooperative Sensing Among Vehicles

A major concern about centralized sensing in CR-VANETs is discussed in [17]. Vehicles may have different views of spectrum occupancy, especially near the edge of the range of primary systems, and thus it is difficult to set a BS or data fusion center. Instead, in [17], a distributed collaborative sensing scheme is proposed. Vehicles send the message about the belief on the existence of primary users to neighboring vehicles, which is called belief propagation (BP). Upon receiving the belief messages, vehicles combine the belief with their local observation to create new belief messages. After several iterations, each vehicle is envisaged to have a stable belief, and can conduct spectrum sensing accordingly. However, several issues of this work could be further discussed, such as the convergence speed of the iterations, the extent of belier propagation, etc.

Cooperative sensing between selected neighboring vehicles can be more efficient than cooperation among all neighboring vehicles due to less message exchanges. A light-weight cooperative sensing scheme can be seen in [18]. Roads are divided into segments, and vehicles are allowed to gather spectrum information of h segments ahead from vehicles in front, which is a priori spectrum availability detection. Therefore, vehicles can decide the channel to use in advance so that spectrum opportunities can be better utilized.

2.1.2 Dynamic Spectrum Access in CR-VANETs

The sensing results should be utilized to correctly choose the spectrum to access. This can be done through different approaches, which can be categorized into PU protection and QoS support, in terms of the target of the approaches.

2.1.2.1 Spectrum Access Approaches with PU Protection

In these approaches, vehicles access the spectrum with the goal of avoiding harmful interference to licensed system and PUs. In [19], a learning structure is proposed for channel selection in CR-VANETs. PU channel usage is modeled as "ON/OFF" pattern, where in ON period, the channel usage follows "busy/idle" pattern, and in OFF period, the PU does not transmit. The authors claimed that an instant spectrum sensing is difficult to differentiate OFF periods from idle periods, while longer sensing time reduces the utilization of OFF periods. Based on the fact that samples of spectrum usage during the same time slot of days at the same location keep high consistency, a channel selection is proposed jointly considering the past channel selection experience and current channel conditions. Stored channel profiles are used to select good channel candidates to sense and access, avoiding wasting limited sensing time on other channels. In [20], some metrics for dynamic channel selection are proposed and discussed. These metrics include: (1) channel data rate; (2) product of channel utilization and data rate; (3) product of expected OFF period and data rate.

2.1.2.2 Spectrum Access with QoS support

QoS support is important for VANETs, such as the delay constraint for safety applications and bandwidth requirement of nonsafety applications. Therefore, QoS support is a crucial consideration in dynamic spectrum access schemes. In [21], a dynamic spectrum access scheme for vehicle-infrastructure uplink communication is proposed to minimize the energy consumptions as well as guarantee the QoS. It is claimed that energy efficient communication is important for VANETs to save energy and reduce greenhouse gas emission, especially for the electric vehicles. A joint optimization algorithm is proposed to minimize the energy consumption while maintain the throughput requirement with the delay constraint of vehicular communications. In [9], a dynamic channel selection scheme is proposed for vehicle clusters, involving dynamic access to shared-use channels, reservation of exclusive-used channels, and control of cluster size. Shared-use channel, i.e., licensed channels, can be accessed by vehicles in an opportunistic manner, while exclusive-use channels are reserved for vehicle data transmission exclusively, such as the DSRC spectrum band located at 5.9 GHz band. Channel selection is modeled as an optimization problem under the constraints of QoS specifications and PU protection, which is solved by constrained Markov decision process.

2.2 Opportunistic WiFi (Unlicensed Band) for VANETs

WiFi, as a popular wireless broadband access technology operating on the unlicensed spectrum, provides the "last-hundred-meter" backhaul connection to private or public Internet users. Through WiFi, data traffic that is originally targeted for cellular networks can be delivered, which is referred to as *WiFi offloading* of the mobile data. Hereafter, we use the term *WiFi offloading* to represent data transmission through WiFi networks. The advantages of WiFi access can be found in Table 2.1. These advantages make WiFi a cost-effective technology to offload the cellular data traffic and alleviate the congestion of cellular networks. As a matter of fact, WiFi is recognized as one of the primary cellular traffic offloading technologies [24]. WiFi offloading has been extensively investigated for stationary or slow moving users[1] in [24–27]. It is shown that about 65 % of the cellular traffic can be offloaded by merely switching the IP connection from the cellular network to

---

[1]We refer to these users as non-vehicular users.

Table 2.1 The advantages of WiFi access

Advantage	Description
Widely deployed infrastructure	WiFi hotspots are widely deployed in many urban areas. It is shown that WiFi access is available 53 % of the time while walking around popular sites in some large cities [22].
Low cost	WiFi access is often free of charge or inexpensive. For example, KT Corporation in South Korea offers WiFi services with $ 10 a month for unlimited data usage [23].
High availability of user devices	Most of current mobile devices, such as smart phones, tablets, and laptops are equipped with WiFi interfaces.
Efficient data transmission	Currently WiFi technologies (IEEE 802.11 b/g) can provide data rates of up to 54 Mbps. There are new technologies under development or test, e.g., IEEE 802.11 ac/ad, which can provide data transfer at several Gbps.

WiFi networks when WiFi connectivity is available (*termed on-the-spot offloading*). Moreover, a large amount (above 80 %) of data can be offloaded by delaying the data application [25] (*termed delayed offloading*), since the mobile users can wait for WiFi connection and then transmit the data.

For moving vehicles, the feasibility of WiFi for outdoor Internet access at vehicular speeds has been demonstrated in [28]. The built-in WiFi radio or WiFi-enabled mobile devices on board can access the Internet when vehicles are moving in the coverage of WiFi hotspots, which is referred to as the **drive-thru Internet** access [29]. This access solution is workable to provide a cost-effective data pipe for VUs [30], and with the increasing deployment of the urban-scale WiFi networks (e.g., Google WiFi in the city of Mountain View) and carrier-WiFi networks (i.e., WiFi networks deployed by cellular carriers), there will be a rapid growth in vehicular Internet connectivity. WiFi offloading in vehicular communication environments (or *vehicular WiFi offloading*) refers to delivering the data to/from VUs through opportunistic WiFi networks, i.e., the drive-thru Internet. Natural questions could arise here. How much data can be offloaded through vehicular WiFi offloading? How to improve the offloading performance, i.e., to offload more cellular traffic and guarantee the QoS of VUs simultaneously? Due to high dynamics of vehicular communication environments, the effectiveness of WiFi offloading for VUs should be carefully studied. The overview of vehicular WiFi offloading is shown in Fig. 2.3. The Unique features and challenges of vehicular WiFi offloading are elaborated from the following three aspects.

Drive-thru Internet access: In vehicular WiFi offloading, mobility plays a both distinguishing and challenging role. During one drive-thru, i.e., the vehicle passing the coverage area of one WiFi AP, the connection time is limited due to the small coverage area of AP and the mobility of VUs, and therefore VUs can only obtain a relatively small volume of data; VUs may experience multiple drive-thrus in a short time period due to high mobility. This short and intermittent connectivity can significantly impact offloading schemes, such as WiFi offloading potential

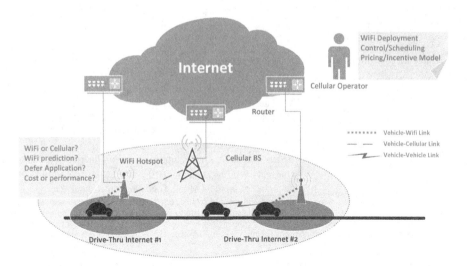

Fig. 2.3 WiFi offloading in vehicular communication environments

prediction and network selection (cellular/WiFi). Fluctuating channels can lead to high and bursty losses, resulting in disruptions to connectivity. Thus, proper handoff schemes and transport protocols should be considered to reduce the disruptions and adapt to the wireless losses.

Cellular operators: To ease congestion in cellular networks, cellular operators may adopt certain commercial strategies to encourage data offloading, one of which is stimulating VUs to transmit their data through opportunistic WiFi networks. Thus, incentive models, such as variable service prices or reward mechanisms, should be investigated. Moreover, cellular operators may deploy their own commercial or non-commercial WiFi networks (carrier-WiFi networks) to offload mobile data, e.g., the WiFi hotspots operated by AT&T [31]. How to determine the WiFi deployment strategy to attain optimal offloading performance is another research challenge.

Vehicular users: As the mobility pattern of vehicles can be partially predicted from the mobility model, historic drive information, and driver preferences, the WiFi offloading potential, i.e., data volume offloaded in the future, can be predicted in a certain level. Based on the prediction and the knowledge of usage cost of cellular and WiFi services, and the QoS requirements, it is possible for VUs to determine when to use WiFi or cellular networks upon a service request emerging, in order to get a good tradeoff between the cost and satisfaction level in terms of delay. It is a challenging task to understand the cost-effectiveness of WiFi offloading from the VUs' perspective.

In this section, we focus on the problem of WiFi offloading in vehicular environment. We discuss the challenges and identify the research issues related to drive-thru Internet as well as vehicular WiFi offloading. Moreover, we review the state-of-the-art solutions, providing rapid access to research results scattered over many papers.

Table 2.2 Drive-thru Internet performance measurements—configuration

	Scenario	WiFi	AP deployment	Antenna
[29]	Highway	802.11b	Planned	External/VU
[33]	Highway	802.11g	Planned	8 dBi/AP; 5 dBi/VU
[34]	Traffic free road	802.11b	Planned	N/A
[35]	Highway	802.11a/b/g	Planned	7 dBi/AP
[28]	Urban	802.11b	Unplanned	5.5 dBi/VU
[32]	Urban	802.11b/g	Unplanned	3 dBi/VU

N/A: not applicable

2.2.1   Drive-thru Internet Access

The performance of drive-thru Internet is different from that of a normal WiFi network which mostly serves non-vehicular users. The reasons are three-fold. Firstly, high vehicle mobility results in a very short connection time to the WiFi AP, e.g., only several to tens of seconds, which greatly limits the volume of data transferred in one connection time. Moreover, the time spent in WiFi association, authentication, and IP configuration before data transfer can take up a considerable part of the short connection time. Secondly, communications in vehicular environments suffer from the high packet loss rate due to the channel fading and shadowing [32]. Thirdly, the stock WiFi protocol stack is not specifically designed for high mobility environments.

Vehicular WiFi offloading mostly relies on the drive-thru Internet access opportunities, provided by open or planed WiFi networks. Therefore, we first review the recent experimental and theoretical studies on drive-thru Internet. After that, we discuss the vehicular WiFi offloading, including the challenges, research issues, and existing and potential solutions.

We elaborate our discussion on drive-thru Internet from the following aspects.

2.2.1.1   Characteristics and Performance

To characterize and evaluate the performance of the drive-thru Internet access, several real-world measurements have been conducted based on diverse test bed experiments. The configurations and key results are summarized in Tables 2.2 and 2.3.

In [29, 33], the performance of the drive-thru Internet access is evaluated in a planned scenario. Two APs are deployed closely along a highway, using IEEE 802.11b and 802.11g, respectively. The performances of User Datagram Protocol (UDP) and Transmission Control Protocol (TCP) at different vehicle speeds (80, 120, and 180 km/h) and scenarios (AP to vehicle, vehicle to AP) are measured. A very important feature of the drive-thru Internet observed from the experiment is that VUs may encounter three phases successively during the drive-thru, i.e., entry,

Table 2.3 Drive-thru Internet performance measurements—results

	Connection establishment time	Connection time	Inter-connection time	Max rate	Data transfer in once drive-thru
[29]	Max 2.5 s	9 s @ 80	N/A	TCP: 4.5 Mbps	TCP: 6 MB @ 80
					5 MB @ 120
					1.5 MB @ 180
				UDP: 5 Mbps	UDP: 8.8 MB @ 80
					7.8 MB @ 120
					2.7 MB @ 180
[33]	N/A	N/A	N/A	15 Mbps	Max 110 MB
[34]	8 s	217 s @ 8	N/A	TCP: 5.5 Mbps	92 MB @ 8
		13.7 s @ 120		UDP: 3.5 Mbps	6.5 MB @ 120
[35]	Mean 13.1 s	58 s	N/A	TCP: 27 Mbps	Median 32 MB
[28]	366 ms	13 s	Mean 75 s	30 KBps	Median 216 KB
[32]	8 s	N/A	Median 32 s	86 Kbps;	Median 32 MB
			Mean 126 s		

@ α: at α (km/h); N/A: not applicable

production, and exit phases. In the entry and exit phases, due to the weak signal, connection establishment delay, rate overestimation, etc, the data transmission performance is not as good as that in the production phase. In [34], a similar test is conducted on a traffic free road which means there is no interference or contention among different VUs. It is shown that in such an environment, the performance of the drive-thru Internet suffers most from the backhaul network or application related issues rather than the wireless link problems. For example, with a 1 Mbps bandwidth limitation of backhaul network, the TCP bulk data transferred within a drive-thru reduces from 92 to 25 MB. In addition, a backhaul with 100 ms one-way delay greatly degrades the performance of web services due to the time penalty of HTTP requests and responses. The discussion on the problems that may cause the performance degradation of the drive-thru Internet can be seen in [35].

In [28, 32], large-scale experimental evaluations with multiple vehicles in urban scenarios have been conducted. Both of the data sets are collected from the city of Boston with in situ open WiFi APs. TCP upload performance is investigated in [28]. It is indicated that with fixed 1 Mbps MAC bit rate, the drive-thru Internet is able to provide an (median) upload throughput of 30 KBps, and the median volume of uploading data in once drive-thru is 216 KB. The average connection and inter-connection (between successive connections) time are 13 s and 75 s, respectively. This shows that although vehicles have short connection time with WiFi APs, they may experience drive-thru access opportunities more frequently, compared with low-mobility scenarios (median connection and inter-connection time 7.4 min and 10.5 min, respectively [25]). In [32], the experiment shows a 86 kbps long-term average data transfer rate averaged over both connection and inter-connection periods. More importantly, two mechanisms to improve the performance are proposed, namely Quick WiFi and CTP, to reduce the connection establishment time and deal with the negative impact of packet loss on transportation layer protocols, respectively.

2.2.1.2 Network Protocol

To improve the performance of the drive-thru Internet, in the literature, new protocols or modification in existing protocols are developed. The efforts in the literature include: (1) improving transport protocols to deal with the intermittent connectivity and wireless losses [32]; (2) reducing connection establishment time [32]; (3) enhancing MAC protocols for high mobility scenarios [36]; and (4) MAC rate selection schemes [32, 35].

To deal with the bursty and high non-congestion wireless losses in vehicular communication environments, a transport protocol called Cabernet transport protocol (CTP) is proposed in [32]. In CTP, a network-independent identifier is used by both the host and the VU, allowing seamless migration among APs. Large send and receive buffers can help to counter the outages (i.e., during the inter-connection time). More importantly, CTP can distinguish wireless losses from congestion losses, by periodically sending probe packets. Through an experimental evaluation, CTP is demonstrated to achieve twice the throughput of TCP in the drive-thru environment.

The connection time between a moving vehicle and a WiFi AP typically ranges from seconds to tens of seconds in drive-thru scenarios, and not all of it can be used for real data transfer. It takes some time to conduct AP association, authentication, IP configuration, etc, before Internet connectivity is available. This time is called connection establishment time. It is straightforward that the performance of data transmission can be improved if this time duration can be reduced. In [32], a mechanism named Quick WiFi is proposed to reduce the connection establishment time and improve data transfer performance. The main idea of Quick WiFi is to incorporate all processes related to connection establishment into one process, to reduce the timeouts of related processes, and to make use of parallelism as much as possible. It is shown that the connection establishment time can be reduced to less than 400 ms. If the WiFi network is deployed and managed by one mobile network operator (MNO), a simple yet effective method to reduce overhead due to connection establishment is presented in [37], in which vehicles are allowed to retain their IP address among different associations, and thus the authentication and IP configuration are carried out only once.

The IEEE 802.11 Wireless Local Area Network (WLAN) MAC protocols are designed for low-mobility scenarios, and consequently require modifications and redesigns for the drive-thru Internet. In [36], the performance of IEEE 802.11 distributed coordination function (DCF) of the large-scale drive-thru Internet is theoretically studied based on a Markov chain model. The impact of vehicle mobility and network size (i.e., vehicle traffic density) on the MAC throughput performance is also discussed. The key observation is that the normal operations

Table 2.4 Modified MAC bit
rate selection scheme in [35]

Parameter	Original [38]	New value
Probe packet	10 %	40 %
Sample window	10 s	1 s
Decision interval	Every 1000 ms	Every 100 ms

of DCF result in *performance anomaly*, i.e., the system performance is deteriorated by the users with lower transmission rates. Based on this observation and the analytical model, a contention window optimization is proposed, which is adaptive to variations of vehicle velocity, transmission rates, and network size. The MAC rate selection methods are discussed in [32, 35]. In [32], a fixed 11 Mbps IEEE 802.11b bit rate is selected for the drive-thru upload scenario due to the following observations: (a) the loss rates for IEEE 802.11b bit rates (1, 2, 5.5, 11 Mbps) are similar; and (b) the IEEE 802.11g bit rates, though may be higher (up to 54 Mbps), all suffer from high loss rate (about 80 %). In [35], it is observed that the original WiFi bit rate selection algorithm is not responsive enough to the vehicular communication environments, and higher rates are rarely selected. By a simple modification shown in Table 2.4, 75 % improvement of TCP goodput can be achieved. However, an optimal MAC bit rate selection scheme which is suitable for the drive-thru Internet is still missing, and is a challenging research task.

2.2.2   Vehicular WiFi Offloading

The roadside WiFi network and vehicles with high mobility constitute a practical solution to offload cellular data traffic, namely, vehicular WiFi offloading. Vehicular WiFi offloading is conducted by multiple drive-thru Internet access networks, and thus has all the features of drive-thru Internet. Vehicular WiFi offloading also has its own features. The research issues in the literature mainly focus on strategies to improve the offloading performance, especially for non-interactive applications which can tolerate certain delays.

WiFi offloading schemes for non-vehicular users often focus on the availability and offloading performance of the current AP, or at most the forthcoming AP, since the user is expected to have a relatively stable connection with one AP. This, however, is not the case in vehicular communication environments. Since VUs may meet multiple APs with different connection qualities within a short time period, offloading schemes should incorporate the prediction of WiFi availability to better exploit multiple data transfer opportunities. Characteristics of data application, e.g., delay and throughput requirements, can also have great impacts on offloading schemes. For example, non-interactive applications, such as email attachments, bulk data transfer, and regular sensing data upload, are often throughput-sensitive, whereas the delay requirements are not very stringent generally. For such applications, a good offloading scheme is to use WiFi as much as possible while guaranteeing that the delay requirement is not violated. On the other hand, interactive applications, such as VoIP and video streaming, are typically delay-sensitive, and it is more challenging to design proper offloading schemes for such applications. We review the literature on vehicular WiFi offloading as follows.

Vehicular WiFi offloading schemes have been proposed in the literature [39–41]. In [39], an offloading scheme named **Wiffler** is proposed to determine whether to defer data applications for the WiFi connectivity or to use cellular networks immediately. Wiffler incorporates the prediction of the WiFi offloading potential along vehicles' route and considers delay requirements of different types of applications. An experiment is first conducted to study the availability and performance characteristics of WiFi and 3G cellular networks, and shows that at more than half of the locations in the target city, at least 20 % of cellular traffic can be offloaded through drive-thru WiFi networks, although the WiFi temporal availability is low (12 % of the time) due to the vehicle mobility. Wiffler enables the delayed offloading and switch to 3G quickly for delay-sensitive applications. The delay requirement of data applications is determined according to VUs' preference or inferred from the application port information or binary names. The effective WiFi throughput, i.e., the volume of data handled through WiFi APs before the maximum delay is expired, can be predicted by estimating the number of encountered APs. For the bursty AP encounters, the prediction is done assuming the inter-contact time durations in the future can be estimated by history average value. Using such a prediction, the traffic is offloaded to WiFi networks when $W > S \cdot c$, where W is the predicted effective WiFi throughput, S is the data size required to be transferred within the delay, and c is called *conservative quotient* to control the tradeoff between the offloading effectiveness and the application completion time. For delay-sensitive applications, a fast switching technology is used when the WiFi link-layer fails to deliver a packet within a predefined time threshold.

Motivated by the fact that the mobility and connectivity of vehicles can be predicted, a pre-fetching mechanism is proposed in [40], in which APs along the predicted vehicle route cache the contents and deliver them to VUs when possible. This can benefit the WiFi offloading since the vehicle-to-AP bandwidth is often higher than the backhaul bandwidth, and vehicle-to-AP transmissions can use specialized transport protocols which is designed to be against the impact of wireless losses (e.g., CTP which is discussed in Sect. 2.2.1.2). The pre-fetching is based on the mobility prediction, and to deal with the impact of prediction errors, the data is allowed to be redundantly pre-fetched by subsequent APs. However, since the WiFi backhaul capability has been greatly enhanced in recent years, the advantage of the pre-fetching solution might be reduced.

In [41], a vehicular WiFi offloading scheme called oSCTP is proposed from the perspective of transport layer to offload the cellular traffic via WiFi networks and maximize the user's benefit. The philosophy of oSCTP is to use WiFi and cellular interfaces simultaneously if necessary, and schedule packets transmitted in each interface every scheduling interval. By modeling user utility and cost as a function of the cellular and WiFi network usage, the user's benefit, i.e., the difference between the utility and the cost, is maximized through an optimization problem.

The experimental evaluation shows a 63–81 % offloaded traffic by using oSCTP, verifying the effectiveness of the proposed offloading scheme.

It has been shown that merely using in situ WiFi APs can hardly provide performance guarantee of the drive-thru Internet. To achieve a satisfied and stable offloading performance, planned WiFi deployment should be considered and in fact is already an ongoing effort. Since it is cost prohibitive to provide a ubiquitous WiFi coverage, how to deploy a set of WiFi APs to provide a better WiFi availability for vehicles has received much research attention. For example, WiFi deployment strategies in vehicular communication environments are studied in [42]. A notion called *alpha coverage* is introduced, which is defined as the distance or expected delay between two successive AP contacts experienced by moving vehicles on the road. The concept of *alpha coverage* is to guarantee the worse-case coverage of WiFi networks. The metric α is defined in the following way: a WiFi deployment that provides α-Coverage guarantees that there is at least one AP on any path which is of length of at least α. Utilizing such a metric to evaluate the AP deployment is reasonable. First, the delay in vehicular WiFi access is typically caused by intermittent connectivity; and second, with such a delay bound, the WiFi offloading potential can be well predicted given the connection time and throughput statistics of one AP connection. As discussed above, predictions of WiFi availability play a vital role in the design of offloading schemes, since such knowledge can facilitate to determine whether and how much to defer applications. Algorithms to achieve budgeted alpha coverage, that is, to find a set of deployment locations of a bounded number of APs to provide the alpha coverage with minimum α, are also proposed and evaluated. One limitation of [42] is that the AP deployment is sparse so that the AP coverage radius is negligible compared to the distance between neighboring APs. This is not the case for urban scenarios where a cluster of dense APs may a small area. Optimizing the WiFi deployment for urban scenarios with diverse vehicle densities is a demanding task.

To sum up, the drive-thru Internet access has been experimentally evaluated, and several WiFi offloading schemes for VUs have been proposed in the literature to improve the offloading performance. However, there is no related work that can theoretically answer the following questions: (1) how much can WiFi offload in vehicular environment; (2) if the VUs can tolerate certain delay, what is the relation between the offloading performance and the delay tolerance. A theoretical analysis of the performance of vehicular WiFi offloading is of great importance for the following reasons. First, offloading performance of scenarios with different AP deployment and vehicle mobility model can be theoretically analyzed, without the time consuming simulation or even complicated and pricey field test. Second, it can provide network operators with some guidance on WiFi deployment (e.g., the density of WiFi APs), pricing strategies, and incentive mechanisms. Finally, for VUs, the theoretical relation between offloading effectiveness (how much Internet access cost can be saved) and average service delay (how much service degradation the user is willing to tolerate) can provide guidelines for efficient offloading decision making.

2.3 Device-to-Device Communication

D2D communication for VANETs is not well studied in the literature, and thus in this section we generally discuss the research issues and existing solutions in D2D communication. In [43], D2D communication was first proposed to enable multi-hop communication in the cellular network. In [44], 3GPP is investigating D2D communications as Proximity Services (ProSe), and proposed tens of use cases, including general use cases and public safety use cases. In addition, The applications and use cases of D2D communication are summarized in [45], including video dissemination [46], machine-to-machine communication [47], multicasting [48], cellular offloading [49], etc. Although D2D communication may be similar to Ad hoc networks and CRN, the main difference is D2D communication often involves the cellular network in the control panel. In terms of the spectrum resources used, D2D communication can be categorized into inband underlay D2D where D2D and cellular communication share the same spectrum resource [50, 51], inband overlay D2D where part of cellular resources are dedicated to D2D communications [52], and outband D2D where D2D communication is carried out over ISM spectrum [53]. Among the three, inband underlay D2D is studied by the majority of the literature. It is more spectral efficient than inband overlay D2D since the spectrum resources are reused. On the other hand, inband D2D can provide guaranteed QoS where outband D2D normally cannot due to the uncontrolled nature of unlicensed spectrum. In the following, we focus on the literature survey of inband underlay D2D communication.

2.3.1 Spectrum Efficiency

Using inband underlay D2D communications, the cellular spectrum efficiency can be increased by exploiting spatial diversity. To achieve this, a series of issues should be addressed, such as interference management, mode selection, resource allocation, etc.

In [54], the authors propose to use the cellular uplink for D2D communications. Received downlink signal power is used for D2D users to determine their pathloss to the cellular base station. Then based on the pathloss, D2D users will adjust their transmit power so that they can communication directly with each other during the uplink frame without causing much interference to the base station. In [50], the mutual interference between cellular and D2D subsystem is addressed when cellular uplink resource is reused for D2D transmissions. Specifically, two interference avoidance mechanisms are proposed to address D2D-cellular and cellular-D2D interference, namely tolerable interference broadcasting approach and interference tracing approach, respectively. In tolerable interference broadcasting approach, the cellular BS calculates and updates the tolerable D2D interference level of each resource unit and broadcasts the table that contains such information. Upon receiving the table, each D2D user equipments (UEs) can calculate the expected interference due to the pathloss to BS and the transmit power. The D2D UEs will prefer to use the resource unit where the expected interference is much lower than tolerable D2D interference level. In interference tracing approach, D2D UEs measure the interference caused by cellular UEs in a certain period and record the average interference. This information is utilized by D2D UEs when choosing communication resource unit to avoid harmful interference from cellular UEs.

In [55], the authors formulate the interference relationships between D2D and cellular links by an interference-aware graph, and propose a near optimal resource allocation algorithm accordingly to obtain the resource assignment solutions at the BS with low computation complexity. In the interference-aware graph, each vertex represents a cellular or D2D link, and each edge has a weight indicating the potential mutual interference between the two vertices. Simulation results show that the proposed algorithm can achieve the sum rate close to the optimal resource sharing scheme. In [56], a network-assisted method is proposed to intelligently manage resources of devices. Resource units and power are jointly allocated to guarantee the signal quality of all users. In [57], the cellular BS monitors the common control channel and broadcasts the allocated resources. Based on the information, D2D users perform radio resource management to avoid interference to cellular UEs.

2.3.2 Power Efficiency

Power efficiency is very crucial in wireless networks because a higher power efficiency can reduce the power consumption of mobile devices, improve the battery lifetime, and reduce the greenhouse gas emission. Power efficiency enhancement is also a very interesting topic in D2D underlaying cellular networks. In [58], a power optimization scheme with joint resource allocation and mode selection is proposed for OFDMA system with D2D communication integrated. The proposed heuristic algorithm first adopts existing subcarrier allocation and adaptive modulation to allocate subcarriers and bits, and then selects transmission mode for each D2D pair, where if the required transmit power of D2D transmission is higher than a certain threshold, cellular mode is employed to avoid harmful interference. Through simulation, it is shown that 20 % power can be saved compared to the traditional OFDMA system without D2D communication.

In [59], a power-efficient mode selection and power allocation scheme are proposed for D2D underlaying cellular networks. First, the optimal power to achieve the maximum power efficiency for all possible modes of each device is calculated by utilizing the concavity of the upper- and lower-bound of power efficiency. Then, the transmission mode sequence is selected by exhaustive search to achieve the maximal power efficiency. The authors show that the proposed joint power allocation and mode selection scheme can perform close to the upper bound in terms of power efficiency. In [60], a joint mode selection, scheduling and power control task for D2D underlaying cellular networks is formulated as an optimization problem and solved. A centralized optimal framework is proposed with the assumption of the availability of a central entity. Then, a sub-optimal distributed with low computational complexity is developed. Via the simulation, the authors show that the proposed method can achieve significant gain of power efficiency over traditional cellular network when the D2D communication distance is no longer than 150 m.

References

1. Zhang N, Lu N, Lu R, Mark J, Shen X (2012) Energy-efficient and trust-aware cooperation in cognitive radio networks. In: Proceedings of IEEE ICC, Ottawa, June 2012
2. Haykin S (2005) Cognitive radio: brain-empowered wireless communications. IEEE J Sel Areas Commun 23(2):201–220
3. Ian F, Won Y, Kaushik R (2009) Crahns: cognitive radio ad hoc networks. Ad Hoc Netw 7(3):810–836
4. Yucek T, Arslan H (2009) A survey of spectrum sensing algorithms for cognitive radio applications. IEEE Commun Surv Tutorials 11(1):116–130
5. De Domenico A, Strinati EC, Di Benedetto M (2012) A survey on mac strategies for cognitive radio networks. IEEE Commun Surv Tutorials 14(1):21–44
6. Sum CS, Villardi G, Rahman M, Baykas T, Tran HN, Lan Z, Sun C, Alemseged Y, Wang J, Song C, woo Pyo C, Filin S, Harada H (2013) Cognitive communication in tv white spaces: an overview of regulations, standards, and technology [accepted from open call]. IEEE Commun Mag 51(7):138–145
7. Kim W, Gerla M, Oh S, Lee K, Kassler A (2011) Coroute: a new cognitive anypath vehicular routing protocol. Wirel Commun Mob Comput 11(12):1588–1602
8. Di Felice M, Doost-Mohammady R, Chowdhury K, Bononi L (2012) Smart radios for smart vehicles: cognitive vehicular networks. IEEE Veh Technol Mag 7(2):26–33
9. Niyato D, Hossain E, Wang P (2011) Optimal channel access management with qos support for cognitive vehicular networks. IEEE Trans Mob Comput 10(5):573–591
10. Pan M, Li P, Fang Y (2012) Cooperative communication aware link scheduling for cognitive vehicular networks. IEEE J Sel Areas Commun 30(4):760–768
11. FCC, FCC press release (2011) [Online]. Available: http://www.fcc.gov/DailyReleases/DailyBusiness/2011/db0126/DA-11-131A1.pdf
12. Al-Ali A, Sun Y, DiFelice M, Paavola J, Chowdhury K (2014) Accessing spectrum databases using interference alignment in vehicular cognitive radio networks. IEEE Trans Veh Technol 64(1):263–272
13. Tv fool coverage maps (2009) [Online]. Available: http://www.tvfool.com/
14. Doost-Mohammady R, Chowdhury KR (2012) Design of spectrum database assisted cognitive radio vehicular networks. In: Proceedings of IEEE CROWNCOM, Stockholm, June 2012, pp 1–5
15. Pagadarai S, Wyglinski AM, Vuyyuru R (2009) Characterization of vacant UHF TV channels for vehicular dynamic spectrum access. In: Proceedings of IEEE VNC, Tokyo, October 2009, pp 1–8
16. Wang X, Ho P (2010) A novel sensing coordination framework for cr-vanets. IEEE Trans Veh Technol 59(4):1936–1948
17. Li H, Irick DK (2010) Collaborative spectrum sensing in cognitive radio vehicular ad hoc networks: belief propagation on highway. In: Proceedings of IEEE VTC-Spring, Taipei, May 2010
18. Marco D, Kaushik R, Luciano B (2010) Analyzing the potential of cooperative cognitive radio technology on inter-vehicle communication. In: Proceedings of the IEEE Wireless Day, Venice,

November 2010

19. Chen S, Vuyyuru R, Altintas O, Wyglinski AM (2011) On optimizing vehicular dynamic spectrum access networks: automation and learning in mobile wireless environments. In: Proceedings of IEEE VNC, Amsterdam, pp 39–46

20. Tsukamoto K, Omori Y, Altintas O, Tsuru M, Oie Y (2009) On spatially-aware channel selection in dynamic spectrum access multi-hop inter-vehicle communications. In: Proceedings of IEEE VTC-Fall, Anchorage, September 2009, pp 1–7

21. Yang C, Fu Y, Zhang Y, Xie S, Yu R (2013) Energy-efficient hybrid spectrum access scheme in cognitive vehicular ad hoc networks. IEEE Commun Lett 17(2):329–332

22. Go Y, Moon Y, Nam G, Park K (2012) A disruption-tolerant transmission protocol for practical mobile data offloading. In: Proceedings of ACM MobiOpp, Zurich, March 2012

23. Choi Y, Ji HW, Park Jy, Kim Hc, Silvester JA (2011) A 3W network strategy for mobile data traffic offloading. IEEE Commun Mag 49(10):118–123

24. Aijaz A, Aghvami H, Amani M (2013) A survey on mobile data offloading: technical and business perspectives. IEEE Wireless Commun 20(2):104–112

25. Lee K, Lee J, Yi Y, Rhee I, Chong S (2013) Mobile data offloading: how much can WiFi deliver? IEEE/ACM Trans Netw 21(2):536–550

26. Singh S, Dhillon H, Andrews J (2012) Offloading in heterogeneous networks: Modeling, analysis, and design insights. IEEE Trans Wireless Commun 12(5):2484–2497

27. Dimatteo S, Hui P, Han B, Li VO (2011) Cellular traffic offloading through WiFi networks. In: Proceedings of IEEE MASS, Valencia, October 2011

28. Bychkovsky V, Hull B, Miu A, Balakrishnan H, Madden S (2006) A measurement study of vehicular internet access using in situ Wi-Fi networks. In: Proceedings of ACM MobiCom, USA, September 2006

29. Ott J, Kutscher D (2004) Drive-thru internet: IEEE 802.11 b for automobile. In: Proceedings of IEEE INFOCOM, China, March 2004

30. Lu N, Zhang N, Cheng N, Shen X, Mark JW, Bai F (2013) Vehicles meet infrastructure: towards capacity-cost tradeoffs for vehicular access networks. IEEE Trans Intell Transp Syst 14(3):1266–1277

31. List of US cities and counties with large scale WiFi networks (2010) [Online]. Available: http://www.scribd.com/doc/32622309/7-June-2010-List-of-Muni-Wifi-Cities

32. Eriksson J, Balakrishnan H, Madden S (2008) Cabernet: vehicular content delivery using WiFi. In: Proceedings of ACM MobiCom, San Francisco, September 2008

33. Ott J, Kutscher D (2004) The "Drive-thru" architecture: WLAN-based internet access on the road. In: Proceedings of IEEE VTC, Milan, May 2004

34. Gass R, Scott J, Diot C (2006) Measurements of in-motion 802.11 networking. In: Proceedings of IEEE WMCSA, Washington, April 2006

35. Hadaller D, Keshav S, Brecht T, Agarwal S (2007) Vehicular opportunistic communication under the microscope. In: Proceedings of ACM MobiSys, San Juan, Puerto Rico, June 2007

36. Luan TH, Ling X, Shen X (2012) MAC in motion: impact of mobility on the MAC of drive-thru internet. IEEE Trans Mob Comput 11(2):305–319

37. Deshpande P, Hou X, Das SR (2010) Performance comparison of 3G and metro-scale WiFi for vehicular network access. In: Proceedings of ACM SIGCOMM internet measurement conference, Melbourne, November 2010

38. Multiband Atheros Driver for WIFI [Online]. Available: http://www.madwifi.org/

39. Balasubramanian A, Mahajan R, Venkataramani A (2010) Augmenting mobile 3G using WiFi. In: Proceedings of ACM MobiSys, USA, June 2010

40. Deshpande P, Kashyap A, Sung C, Das SR (2009) Predictive methods for improved vehicular WiFi access. In: Proceedings of ACM MobiSys, Kraków, June 2009

41. Hou X, Deshpande P, Das SR (2011) Moving bits from 3G to metro-scale WiFi for vehicular network access: an integrated transport layer solution. In: Proceedings of IEEE ICNP, Vancouver, October 2011

42. Zheng Z, Sinha P, Kumar S (2012) Sparse WiFi deployment for vehicular internet access with bounded interconnection gap. IEEE/ACM Trans Netw 20(3):956–969

43. Lin YD, Hsu YC (2000) Multihop cellular: a new architecture for wireless communications. In: Proceedings of IEEE INFOCOM, Tel Aviv, March 2000, pp 1273–1282

44. 3GPP (2012) Feasibility study for Proximity Services (ProSe) (Release 12), v. 12.2.0, June 2012

45. Asadi A, Wang Q, Mancuso V (2014) A survey on device-to-device communication in cellular networks. IEEE Commun Surv Tutorials 16(4):1801–1819

46. Doppler K, Rinne M, Wijting C, Ribeiro C, Hugl K (2009) Device-to-device communication as an underlay to lte-advanced networks. IEEE Commun Mag 47(12):42–49

47. Zheng K, Hu F, Wang W, Xiang W, Dohler M (2012) Radio resource allocation in LTE-advanced cellular networks with M2M communications. IEEE Commun Mag 50(7):184–192

48. Zhou B, Hu H, Huang S-Q, Chen H-H (2013) Intracluster device-to-device relay algorithm with optimal resource utilization. IEEE Trans Veh Technol 62(5):2315–2326

49. Bao X, Lee U, Rimac I, Choudhury RR (2010) DataSpotting: offloading cellular traffic via managed device-to-device data transfer at data spots. ACM SIGMOBILE Mob Comput Commun Rev 14(3):37–39

50. Peng T, Lu Q, Wang H, Xu S, Wang W (2009) Interference avoidance mechanisms in the hybrid cellular and device-to-device systems. In: Proceedings of IEEE PIMRC, Tokyo, August 2009, pp 617–621

51. Lei L, Zhong Z, Lin C, Shen X (2012) Operator controlled device-to-device communications in LTE-advanced Networks. IEEE Wireless Commun 19(3):96–104

52. Pei Y, Liang YC (2013) Resource allocation for device-to-device communications overlaying two-way cellular networks. IEEE Trans Wireless Commun 12(7):3611–3621

53. Asadi A, Mancuso V (2013) On the compound impact of opportunistic scheduling and d2d communications in cellular networks. In: Proceedings of ACM MSWiM, Barcelona, November 2013, pp 279–288

54. Kaufman B, Aazhang B (2008) Cellular networks with an overlaid device to device network. In: Proceedings of Asilomar conference on signals, systems and computers, Pacific Grove, October 2008, pp 1537–1541

55. Zhang R, Cheng X, Yang L, Jiao B (2013) Interference-aware graph based resource sharing for device-to-device communications underlaying cellular networks. In: Proceedings of IEEE WCNC, Shanghai, April 2013, pp 140–145

56. Tsai AH, Wang LC, Huang JH, Lin TM (2012) Intelligent resource management for device-to-device (D2D) communications in heterogeneous networks. In: Proceedings of IEEE WPMC, China, September 2012

57. Xu S, Wang H, Chen T, Huang Q, Peng T (2010) Effective interference cancellation scheme for device-to-device communication underlaying cellular networks. In: Proceedings of IEEE VTC, Ottwa, September 2010

58. Xiao X, Tao X, Lu J (2011) A qos-aware power optimization scheme in ofdma systems with integrated device-to-device (d2d) communications. In: Proceedings of IEEE VTC-fall, San Francisco, September 2011, pp 1–5

59. Jung M, Hwang K, Choi S (2012) Joint mode selection and power allocation scheme for power-efficient device-to-device (D2D) communication. In: Proceedings of IEEE VTC-Spring, Yokohama, May 2012

60. Belleschi M, Fodor G, Abrardo A (2011) Performance analysis of a distributed resource allocation scheme for d2d communications. In: Proceedings of IEEE GLOBECOM, Houston, December 2011, pp 358–362

Chapter 3
Opportunistic Spectrum Access Through Cognitive Radio

As mentioned above, CR is envisioned as a promising approach to deal with the spectrum scarcity in wireless communications, which enables unlicensed users to opportunistically exploit the spectrum owned by licensed users [1, 2]. In CR-VANETs, considering the highly dynamic mobility, VUs are expected to exploit more spatial and temporal spectrum opportunities along the road than stationary SUs.

In this chapter, we first study the channel availability for CR-VANETs in urban scenarios, taking the mobility pattern of vehicles into consideration. Exploiting the statistics of the channel availability, a distributed opportunistic spectrum access scheme based on a non-cooperative congestion game is proposed for vehicles to exploit spatial and temporal access opportunities of the licensed spectrum. Specifically, we consider a grid-like urban street pattern which models the downtown area of a city. Vehicles equipped with a cognitive radio, moving in the grid, opportunistically access the spectrum of the primary network. The probability distribution of the channel availability is obtained by means of a continuous-time Markov chain (CTMC). After that, we employ a non-cooperative congestion game to solve the problem of vehicles accessing multiple channels with different channel availabilities. We prove the existence of the pure Nash equilibrium (NE) and analyze the efficiency of different NEs, when applying uniform MAC and slotted ALOHA, respectively. A distributed spectrum access algorithm is then developed for vehicles to choose an access channel in a distributed manner, so that a pure NE with high efficiency and fairness is achieved. Finally, simulation results validate our analysis and demonstrate that, with the proposed spectrum access scheme, vehicles can achieve higher utility and fairness compared with the random access. The complete results of this chapter can be seen in [3].

The remainder of the chapter is organized as follows. The detailed description of the system model is provided in Sect. 3.1. In Sect. 3.2, channel availability is analyzed for CR-VANETs in urban scenarios. A spectrum access scheme based

© The Author(s) 2016
N. Cheng, X. (Sherman) Shen, *Opportunistic Spectrum Utilization in Vehicular Communication Networks*, SpringerBriefs in Electrical and Computer Engineering, DOI 10.1007/978-3-319-20445-1_3

on game theory is presented in Sect. 3.3. Simulation results are given in Sect. 3.4. Section 3.5 concludes the chapter.

3.1 System Model

We consider urban scenarios of CR-VANETs, where the transmitters of the primary network are referred to as primary transmitters (PTs), such as TV broadcasters and cellular base stations. Vehicles equipped with a cognitive radio can opportunistically access the licensed spectrum as SUs. There is a non-empty set $\mathcal{K}$ of licensed channels that can be accessed by VUs. The channel usage behavior of PTs and vehicle mobility lead to intermittent channel availability for vehicle. The spectrum opportunity is characterized by *channel availability* experienced by a VU, which is defined as the length of time durations in which the channel is available or unavailable for that VU. The availability of channel $i, i \in \mathcal{K}$, for a VU is determined by the locations and the temporal channel usage pattern of PTs that operate on channel i, as well as the mobility of the vehicle.

3.1.1 Urban Street Pattern

A grid-like street layout is considered for analyzing CR-VANETs in urban environments, like the downtown area of many cities, such as Houston and Portland [4]. The network geometry comprises of a set of vertical roads intersected with another set of horizontal roads. As shown in Fig. 3.1, each line segment represents

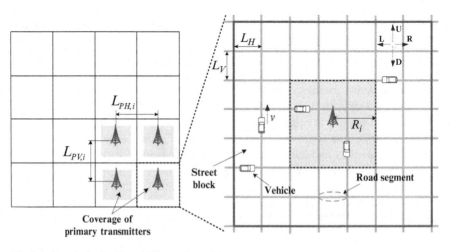

Fig. 3.1 Regularly distributed PTs on channel i

a road segment (the road section between any two neighboring intersections) with bi-directional vehicle traffic. Moreover, all the horizontal segments are considered to have the same length L_H, and all the vertical segments the same length L_V, leading to equal-sized street blocks of $L_H \times L_V$. For example, L_H and L_V are generally from 80 to 200 m for the downtown area of Toronto [5].

3.1.2 Spatial Distribution of PTs

We consider that PTs operating on a generic channel i are regularly distributed in the grid, as shown in Fig. 3.1. The distance between any two neighboring PTs in the horizontal direction and vertical direction is denoted by $L_{PH,i}$ and $L_{PV,i}$, respectively. Denote by R_i the transmission range of PTs on channel i. The coverage area of the PT is approximated by a square area with side length $2R_i$, where $R_i < \min(L_{PH,i}, L_{PV,i})$ to avoid overlapping of different coverage regions of PTs. The approximate coverage area is larger than the real coverage area to protect the primary transmission. A similar approximation of the PT coverage area can be found in [6].

3.1.3 Temporal Channel Usage Pattern of PTs

The temporal channel usage of PTs operating on a generic channel i is modeled as an alternating busy (the PT is active in transmitting) and idle (the PT does not transmit) process [7, 8]. During the transmission period of a PT, vehicles in the coverage area of the PT are not permitted to use the same channel in order to avoid the interference to the primary network. The lengths of busy and idle periods are modeled as exponential random variables with parameters $\lambda_{busy}/\lambda_{idle}$, i.e.,

$$T_{busy,i} \sim Exp(\lambda_{busy,i}) \text{ and } T_{idle,i} \sim Exp(\lambda_{idle,i}), \qquad (3.1)$$

where $X \sim Exp(\lambda)$ indicates that variable X follows an exponential distribution with parameter λ. $\varpi_{idle,i} = \frac{\lambda_{busy,i}}{\lambda_{idle,i} + \lambda_{busy,i}}$ and $\varpi_{busy,i} = 1 - \varpi_{idle,i}$ are the steady-state probabilities that a PT on channel i is active and inactive, respectively.

Note that the spatial and temporal parameters are same for PTs operating on the same channel, but may be different for PTs operating on different channels. For example, if PT1 operates on channel i and PT2 operates on channel j, where $i \neq j, i, j \in \mathscr{C}$, then the transmission range R_1 and R_2 may be different.

3.1.4 Mobility Model

Vehicles move in the grid at a random and slowly changing speed v, where $v \in [v_{min}, v_{max}]$. The average value of v is denoted by $\bar{v}$. At each intersection, vehicles randomly chooses the direction of north, south, east and west with probability P_n, P_s, P_e, and P_w to move on, respectively, as shown in Fig. 3.1. It holds that $P_n + P_s + P_e + P_w = 1$. Once the vehicle chooses a direction at an intersection, it moves straight until it arrives at the next intersection.

3.2 Channel Availability Analysis

The statistics of channel availability can be utilized to design an efficient spectrum access scheme which can improve the QoS of SUs and the spectrum utilization. In this section, we analyze the availability of a specific channel i for VUs, jointly considering the spatial features and the temporal channel usage pattern of PTs, and the mobility of vehicles. It is assumed that PTs operating on the same channel belong to the same type of system and have the same spatial features and temporal channel usage pattern. A similar assumption can be seen in [9]. A continuous-time Markov chain that consists of three states is employed. Denote by S_{Idle}, S_{Busy} and $S_{\overline{C}}$ the states of a VU in the coverage of an idle PT, in the coverage of an busy PT, and outside the coverage of any PT that operates on channel i, respectively, as shown in Fig. 3.2. It can be seen that when the vehicle moves along the street, the state transits to one another. Since the channel is unavailable only when the vehicle is in S_{Busy}, we can further merge S_{Idle} and $S_{\overline{C}}$ as one state in which the channel is available for the VU, denoted by S_A. The state of the channel being unavailable is denoted by S_U, which in fact is S_{Busy}.

Denote the sojourn time of S_A and S_U by $T_{A,i}$ and $T_{U,i}$, respectively. To obtain the probability distribution of channel availability, i.e., the probability distribution of $T_{A,i}$ and $T_{U,i}$, it is necessary to analyze the transition rates among the three states S_{Idle}, S_{Busy} and $S_{\overline{C}}$. Denote by $T_{in,i}$ and $T_{out,i}$ the time duration in which the vehicle remains within the coverage area of a PT and outside the coverage area of any PT on channel i, respectively. Therefore, the transition rates are closely related to the probability distribution of $T_{in,i}$ and $T_{out,i}$. In the following, we focus on the analysis of these two time durations in urban scenarios.

3.2.1 Analysis of T_{in} in Urban Scenarios

To analyze T_{in}, we consider the case in which vehicles move within the coverage of a PT. Denote by Ω_R the coverage area of the PT. Recall that Ω_R is a square

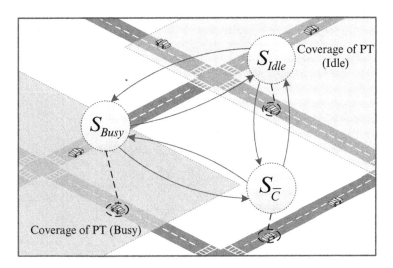

Fig. 3.2 The states of a vehicle w.r.t. the mobility

with each side of length $N_{R,i}$. For ease of the analysis, let $L_V = L_H = L$, and $L_{PH,i} = L_{PV,i} = L_{P,i}$. All lengths are normalized by L, for example, $N_{R,i} = \lceil \frac{2R_i}{L} \rceil$.

In order to analyze T_{in}, a two-dimensional discrete Markov chain is employed, as shown in Fig. 3.3. We index all the intersections within Ω_R, and let each intersection (b, k) be a state denoted by $C_{b,k}$. All these states form a Markov chain

$$\mathbb{C}_{b,k} = \{C_{1,1}, C_{1,2}, \dots, C_{1,N_{R,i}}, C_{2,1}, C_{2,2}, \dots, C_{N_{R,i},N_{R,i}}\}. \quad (3.2)$$

It is said that a vehicle is in state $C_{b,k}$ if it is moving from intersection (b, k) to a neighboring intersection (b', k'). When the vehicle arrives at intersection (b', k'), the state transits from $C_{b,k}$ to $C_{b',k'}$. The states that lie on the boundary of Ω_R are referred to as absorbing states which indicate that the vehicle moves out of the PT coverage area. Let $\mathbf{Q_A}$ be the set of absorbing states. Denote by M the number of transitions it takes before the vehicle leaves Ω_R, i.e., transits to any state in $\mathbf{Q_A}$. T_{in} can be approximated by $M * \Delta t$, where Δt is the time that the vehicle moves through a road segment. To obtain the probability distribution of T_{in}, we should find the probability distribution of M. To this end, we first obtain the transition probabilities of $\mathbb{C}_{b,k}$ as follows:

$$\begin{cases} P(C_{b,k-1}|C_{b,k}) = P_l \\ P(C_{b,k+1}|C_{b,k}) = P_r \\ P(C_{b-1,k}|C_{b,k}) = P_u \quad C_{b,k} \notin \mathbf{Q_A} \\ P(C_{b+1,k}|C_{b,k}) = P_d \\ P(other|C_{b,k}) = 0 \end{cases} \quad (3.3)$$

Fig. 3.3 Analysis of T_{in}: a two-dimensional Markov chain

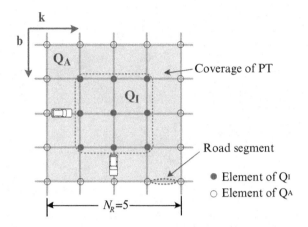

$$\begin{cases} P(C_{b,k}|C_{b,k}) = 1 & C_{b,k} \in \mathbf{Q_A}. \\ P(\text{other}|C_{b,k}) = 0 \end{cases} \qquad (3.4)$$

Denote by $\pi^{(m)}$ the probability distribution of the states after m transitions. Specifically, $\pi^{(0)}$ is the probability distribution of the initial states. It holds that $\pi^{(m)} = \pi^{(0)}\mathbf{P}^m$. At initial time t_0, it is possible for the vehicle to be in any state in Ω_R except those in $\mathbf{Q_A}$. Denote by $\mathbf{Q_I}$ the set of these possible initial states. Then the cardinality of $\mathbf{Q_I}$, denoted by C_I, can be calculated by $C_I = |\mathbf{Q_I}| = (N_{R,i} - 2)^2$. All possible initial states are considered to be with equal probability, and thus $\pi^{(0)}$ can be obtained as follows:

$$\pi^{(0)}_{(b,k)} = \begin{cases} p_I = \frac{1}{C_I} = \frac{1}{(N_{R,i}-2)^2} & C_{b,k} \in \mathbf{Q_I} \\ 0 & \text{otherwise,} \end{cases} \qquad (3.5)$$

where p_I is the probability of each possible initial state. The probability of the event that M is no more than m is given by

$$\mathbf{Pr}(M \leq m) = \sum_{C_{b,k} \in \mathbf{Q_A}} \pi^{(m)}. \qquad (3.6)$$

Therefore, the probability mass function of M is

$$\begin{aligned} \mathbf{Pr}(M = m) &= \mathbf{Pr}(M \leq m) - \mathbf{Pr}(M \leq m-1) \\ &= \sum_{C_{b,k} \in \mathbf{Q_A}} \pi^{(m)} - \sum_{C_{b,k} \in \mathbf{Q_A}} \pi^{(m-1)}. \end{aligned} \qquad (3.7)$$

On the other hand, if all the states in $\mathbf{Q_A}$ are considered as one state S_{End} and all other states as another state S_{Begin}, the two-dimensional Markov chain $\mathbb{C}_{b,k}$ can

Fig. 3.4 Analysis of T_{out}: a two-dimensional Markov chain

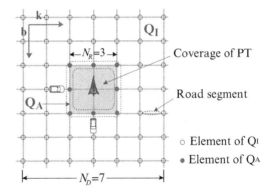

be reduced to a two-state Markov chain $\{S_{Begin}, S_{End}\}$. The vehicle is in S_{Begin} at the beginning. In each transition, it either transits to S_{End} with probability p_0 or remains in S_{Begin} with probability $1 - p_0$, where p_0 is as follows:

$$p_0 = \frac{1}{C_I} \sum_{C_{b,k} \in \mathbf{Q_I}} \sum_{C_{b',k'} \in \mathbf{Q_A}} P(C_{b',k'}|C_{b,k}). \tag{3.8}$$

The state continues to transit until reaching S_{End}. Thus, the number of transitions before the VU leaves the PT coverage area can be considered to follow a geometric distribution with $p = p_0$. From this perspective, T_{in} can be approximated by an exponential distribution, which is discussed later.

3.2.2   Analysis of T_{out} in Urban Scenarios

A two-dimensional Markov chain is employed to analyze T_{out}. Since PTs operating on the same channel are regularly deployed, we can just take the area around one PT to analyze, as shown in Fig. 3.4. Denote this square area by Ω_D, with each side of length $N_{D,i} = \lceil \frac{L_{P,i}}{L} \rceil$. Ω_D is considered to be a torus: when a vehicle leaves the boundary of Ω_D, it moves into Ω_D on the same road from the opposite side of the area. In this situation, the intersections that lie on the boundary of Ω_R are referred to as absorbing states indicating that vehicles in these states move into the coverage of a PT. We can obtain the transition matrix $\mathbf{P}$, which is similar to (3.3) and (3.4) except

$$\begin{cases} P(C_{b,N_D}|C_{b,1}) = P_l \\ P(C_{b,1}|C_{b,N_D}) = P_r \\ P(C_{N_D,k}|C_{1,k}) = P_u \\ P(C_{1,k}|C_{N_D,k}) = P_d. \end{cases} \tag{3.9}$$

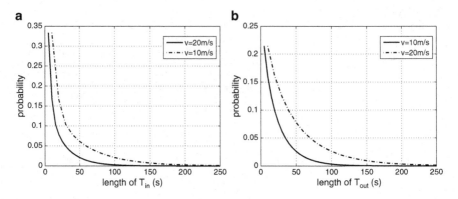

Fig. 3.5 PDF of T_{in} and T_{out} ($L = 100$ m). (**a**) PDF of T_{in}; (**b**) PDF of T_{out}

States which are within Ω_D but outside Ω_R are initial states. Similar to the analysis of T_{in}, the possible initial states are with equal probability. Denote by $\mathbf{Q_I}$ the set of possible initial states, and $|\mathbf{Q_I}| = N_{D,i}^2 - N_{R,i}^2$. Then the probability distribution of initial states, denoted by $\pi^{(0)}$, is as follows:

$$\pi_{(b,k)}^{(0)} = \begin{cases} p_I = \frac{1}{N_{D,i}^2 - N_{R,i}^2} & C_{b,k} \in \mathbf{Q_I} \\ 0 & \text{otherwise.} \end{cases} \quad (3.10)$$

M' denotes the number of transitions before a vehicle moves into the coverage area of a PT. Similar to (3.7), we can get the probability mass function of M' as follows:

$$\begin{aligned} \mathbf{Pr}(M' = m) &= \mathbf{Pr}(M' \le m) - \mathbf{Pr}(M' \le m - 1) \\ &= \sum_{C_{b,k} \in \mathbf{Q_A}} \pi^{(m)} - \sum_{C_{b,k} \in \mathbf{Q_A}} \pi^{(m-1)}. \end{aligned} \quad (3.11)$$

3.2.3 Estimation of λ_{in} and λ_{out}

The probability density function (PDF) of T_{in} and T_{out} from (3.7) and (3.11) are shown in Fig. 3.5. It can be seen that both T_{in} and T_{out} can be approximated by an exponential distribution. Furthermore, the parameter of the distribution can be estimated by using maximum likelihood estimation (MLE) as follows:

$$\lambda = \frac{1}{\bar{x}}, \quad (3.12)$$

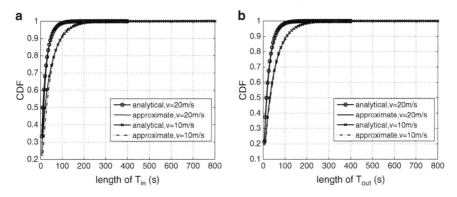

Fig. 3.6 Comparison of analytical and approximate results. (**a**) CDF of T_{in}; (**b**) CDF of T_{out}

where $\bar{x}$ is the sample mean, and λ is the estimated parameter of the exponential distribution. More interestingly, the expected value of T_{in} and T_{out} (denoted by $\overline{T}_{in}$ and $\overline{T}_{out}$, respectively) change with vehicle speed $\bar{v}$, and the spatial parameters of PTs, i.e., R_i and $L_{P,i}$. Specifically, the parameters of the two exponential distributions can be approximated by

$$\lambda_{in} \approx \frac{\bar{v}}{R_i} \text{ and } \lambda_{out} \approx \frac{\bar{v}}{f(L_{P,i} - R_i)}. \tag{3.13}$$

When $L_{p,i} - R_i < 5L$, and $f(\cdot)$ is a linear function, we have

$$\lambda_{out} \approx \frac{\bar{v}}{17.4(L_{P,i} - R_i) - 16.4}. \tag{3.14}$$

Figure 3.6 shows the comparison between analytical and approximate cumulative distribution function (CDF) of T_{in} and T_{out}. It can be seen that they closely match each other, which validates the accuracy of the estimation. The effect of R_i ($N_{R,i}$) and $L_{P,i}$ ($N_{D,i}$) on $\overline{T}_{in}$ and $\overline{T}_{out}$ is shown in Fig. 3.7a, b, respectively. A larger value of R_i leads to a larger value of $\overline{T}_{in}$, while a larger value of $L_{P,i}$ leads to a larger value of $\overline{T}_{out}$, which is consistent with the expectation.

3.2.4 Derivation of Channel Availability

From the above analysis, the transition rates among the states of the Markov chain shown in Fig. 3.2 can be obtained, as listed in Table 3.1. Denote by ζ_i the average

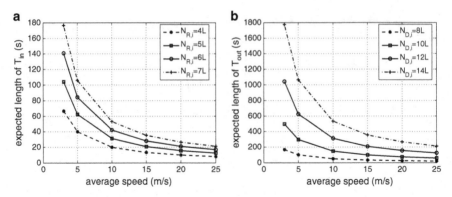

Fig. 3.7 Impact of $N_{R,i}$ and $N_{D,i}$ on $\overline{T}_{in}$ and $\overline{T}_{out}$. (**a**) Impact of $N_{R,i}$ on $\overline{T}_{in}$; (**b**) impact of $N_{D,i}$ on $\overline{T}_{out}$

Table 3.1 State transition rates of a vehicle for channel i

State transition	Rate
$S_{idle} \rightarrow S_{busy}$	$\lambda_{idle,i}$
$S_{busy} \rightarrow S_{idle}$	$\lambda_{busy,i}$
$S_{busy} \rightarrow S_{\overline{C}}$	$\frac{\overline{v}}{R_i}$
$S_{\overline{C}} \rightarrow S_{busy}$	$\frac{\overline{v}}{f(L_{p,i}-R_i)}\varpi_{busy,i}$
$S_{idle} \rightarrow S_{\overline{C}}$	$\frac{\overline{v}}{R_i}$
$S_{\overline{C}} \rightarrow S_{idle}$	$\frac{\overline{v}}{f(L_{p,i}-R_i)}\varpi_{idle,i}$

fraction of the area of PT coverage on channel i, and $\zeta_i = \frac{4R_i^2}{L_{p,i}^2}$. Thus, the average fraction of areas where channel i is available at any given time, denoted by δ_i, can be given by:

$$\delta_i = (1 - \zeta_i) + \zeta_i \varpi_{idle,i} = 1 - \zeta_i \varpi_{busy,i}, \qquad (3.15)$$

The state $S_{U,i}$ ends when the VU moves out of the coverage of the PT or the PT stops transmission. Therefore, $T_{U,i}$ follows an exponential distribution with parameter $\lambda_{U,i}$, where

$$\lambda_{U,i} = \lambda_{busy,i} + \frac{\overline{v}}{f(L_{p,i} - R_i)}\varpi_{busy,i}. \qquad (3.16)$$

Based on the balance condition $\varpi_{A,i}\lambda_{A,i} = \varpi_{U,i}\lambda_{U,i}$, where $\varpi_{A,i} = \delta_i$ and $\varpi_{U,i} = 1 - \delta_i$, we can get the $S_{A,i} \rightarrow S_{U,i}$ transition rate $\lambda_{A,i}$ as follows:

$$\lambda_{A,i} = \frac{\delta_i}{1 - \delta_i}\lambda_{U,i} = \frac{\zeta_i\varpi_{busy,i}}{1 - \zeta_i\varpi_{busy,i}}\left(\lambda_{busy,i} + \frac{\overline{v}\cdot\varpi_{busy,i}}{f(L_{p,i} - R_i)}\right), \qquad (3.17)$$

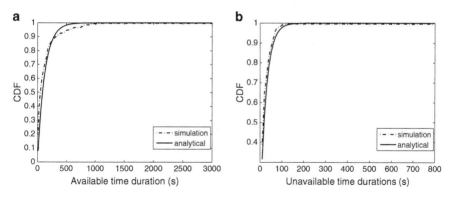

Fig. 3.8 The analytical and simulation results of T_A and T_U. (**a**) T_A: channel available period; (**b**) T_U: channel unavailable period

and thus, $T_{A,i} \sim Exp(\lambda_{A,i})$. Figure 3.8 shows the comparison between analytical and simulation results. The two curves closely match to each other for both $T_{A,i}$ and $T_{U,i}$, which demonstrates the accuracy of the analysis.

Define the effective channel availability (ECA) Ψ of channel i as the average time duration in which channel i is available for a VU to access, which can be calculated as follows:

$$\Psi_i = \eta_i \cdot \overline{T}_{A,i} = \frac{\eta_i}{\lambda_{A,i}}, \tag{3.18}$$

where $\eta_i \in (0, 1)$ is the interference factor representing the tolerance level of interference of primary network. Note that a larger value of η brings more spectrum opportunities, but at the same time leads to more interference to the primary network.

Considering a real-world road map can facilitate a more precise analysis of channel availability. This, however, introduces cumbersome challenges. Our approach is based on a simple regular road pattern, which offers a workable approximation.

3.3 Game Theoretic Spectrum Access Scheme

From the previous section, the channel availability statistics, i.e., ECA of each channel, are obtained. To design a spectrum access scheme, we assume that VUs are aware of the spatial features and temporal channel usage pattern of PTs of each channel, i.e., $L_{PH,i}$, $L_{PV,i}$, R_i, $\lambda_{busy,i}$ and $\lambda_{idle,i}$, since the information of primary networks can be obtained from network operators. With such information, vehicles can obtain the ECA of each channel, i.e., $\Psi_i, i \in \mathcal{K}$, based on their speed, by using (3.17) and (3.18). Before transmitting, VUs conduct spectrum sensing, which is considered accurate in this work. Since the channel availability follows the exponential distribution which is memoryless, the channel availability is independent of the situation before the spectrum sensing. Notably, if a channel is unavailable due to primary transmissions, its ECA is zero. The bandwidth of each channel is considered to be identical. Vehicles on the road often maintain relatively stable topology since they follow the same direction and similar speed, and thus they usually form clusters where vehicles within a cluster can communicate with each other [10, 11]. To utilize the licensed spectrum, VUs should opportunistically access the channels with different ECAs in a distributed manner. Game theory [12] is a well-known tool to analyze the behavior of distributed players who are considered to be selfish and rational. We apply a non-cooperative congestion game to model the spectrum access process in CR-VANETs, in which VUs in a cluster choose channels to access in a distributed manner, trying to maximize their own utility. Then, we analyze the existence and the efficiency of NE using uniform MAC and slotted ALOHA, respectively, and devise a spectrum access algorithm for each vehicle to decide the access channel and meanwhile achieve an NE with high efficiency.

3.3.1 Formulation of Spectrum Access Game

The spectrum access problem is modeled as a congestion game, where there are multiple players and resources, and the payoff of each player by selecting one resource is related to the number of other players selecting the same resource [13]. In this chapter, the spectrum access congestion game is defined as $\Gamma = \{\mathcal{N}, \mathcal{C}, \{S_j\}_{j\in\mathcal{N}}, \{U_j\}_{j\in\mathcal{N}}\}$, where $\mathcal{N} = \{1, \ldots, N\}$ is the finite set of players, i.e., VUs in a cluster. N is related to the vehicle density, which is denoted by ρ_v; $\mathcal{C} = \{1, \ldots, C\}$ is the set of available channels, where "available" means that the channels are sensed to be idle, and $\mathcal{C} \subseteq \mathcal{K}$; S_j is the set of pure strategies associated with VU j; and U_j is the utility function of VU j. VUs are aware of the ECA of all channels (Ψ_i) and the number of VUs in the game (N). The bandwidth (resource) of each channel is identical, and we normalize the bandwidth to one unit.

Each VU can access at most one channel at a time since only one cognitive radio is equipped, and thus $S_j = \mathcal{C}$ for all $j \in \mathcal{N}$. In this case, denote by U_j^i the utility of VU j when channel i is chosen. Note that U_j^i is a function of both s_j and s_{-j}, which are the strategies selected by VU j and all other VUs, respectively. In this game, we define the utility U_j^i as the average total channel resource VU j obtains by choosing channel i, before this channel is reoccupied by PTs, i.e.,

$$U_j^i = \Psi_i r(n_i). \tag{3.19}$$

n_i is the total number of VUs choosing channel i simultaneously, including VU j. Resource allocation function $r(n_i)$ indicates the share of channel i obtained by each of the n_i vehicles. For an arbitrary VU, Ψ_i is used to measure the average time duration in which channel i is available. Thus, $U_j^i = \Psi_i r(n_i)$ gives the average total amount of channel resource VU j can obtain before it must cease transmitting due to the appearance of active PTs. The channel with higher Ψ_i is preferred because choosing it can reduce spectrum sensing and unpredictable channel switching. The form of $r(\cdot)$ is related to specific MAC schemes. However, based on [14], $r(\cdot)$ should satisfy the following conditions:

- $r(1) = 1$, which means that a user can get all the resource of a channel if it is the only one choosing that channel.
- $r(n)$ is a decreasing function of n.
- Define $f(n) = nr(n)$. $f(n)$ decreases with n and should be convex, i.e., $f'(n) < 0$ and $f''(n) > 0$.
- $n_i r(n_i) \leq 1$. Resource waste may happen when multiple users share the same resource due to contention or collision.

Since VUs are rational and selfish, they prefer the strategy that can maximize their utilities. To analyze this game, we analyze the existence, condition and efficiency ratio (ER) of the pure NE, using uniform MAC and slotted ALOHA, respectively. After that, a spectrum access algorithm to achieve the pure NE with high ER is derived.

3.3.2 Nash Equilibrium in Channel Access Game

Nash equilibrium is a well-known concept to analyze the outcome of the game, which states that in the equilibrium every user can select a utility-maximizing strategy given the strategies of other users.

Definition 1. A strategy profile for the players $S^* = (s_1^*, s_2^*, \ldots, s_N^*)$ is an NE if and only if

$$U_j(s_j^*, s_{-j}^*) \geq U_j(s_j', s_{-j}^*), \forall j \in \mathcal{N}, s_j' \in S_j, \tag{3.20}$$

which means that no one can increase his/her utility alone by changing its own strategy, given strategies of the other users. If the strategy profile S in (3.20) is deterministic, it is called a pure NE. In this chapter, we consider pure NE only, and therefore we use the term NE and pure NE interchangeably.

In the spectrum access game Γ, given a strategy profile, if no VU can improve his/her utility by shifting to another channel alone, the strategy profile is referred to as a pure NE. Denote by $S = (s_1, s_2, \ldots, s_N)$ the strategy profile of all VUs, where s_i is a specific selected channel. Denote by $\mathbf{n}(S) = (n_1, n_2, \ldots, n_C)$ the congestion vector, which shows the number of VUs choosing each channel, corresponding to the strategy profile S. According to Definition 1, the spectrum access game Γ has pure NE(s) if and only if for each VU $j \in \mathcal{N}$,

$$\Psi_{s_j} r(n_{s_j}) \geq \Psi_k r(n_k + 1), \forall k \in \mathcal{C}, k \neq s_j. \tag{3.21}$$

Note that there are typically multiple strategy profiles that correspond to one congestion vector. If a strategy profile S corresponding to congestion vector $\mathbf{n}^*$ is a pure NE, then all strategy profiles corresponding to $\mathbf{n}^*$ are pure NEs according to (3.21). Denote by NE-*set*($\mathbf{n}$) the set of pure NEs corresponding to congestion vector $\mathbf{n}$. The NEs in NE-*set*($\mathbf{n}$) may yield different utilities for each player. However, they yield the same total utility, which is defined as the summation of the utilities of all vehicles, and is given by

$$U_{total,\mathbf{n}} = \sum_{i=1}^{C} \Psi_i n_i r(n_i) = \sum_{i=1}^{C} \Psi_i f(n_i), \qquad (3.22)$$

where $\mathbf{n} = (n_1, n_2, \ldots, n_C)$. In the following, the NE of the spectrum access game is analyzed using uniform MAC and slotted ALOHA, respectively.

3.3.3 Uniform MAC

The simplest way to share the channel among multiple users is to make each of them access the channel equally likely, which is referred as to uniform MAC [14]. Each VU starts a back-off with the back-off time randomly chosen from a fixed window. If one VU finds that its back-off expires and the channel is idle, it can capture the channel during the whole time slot, while others should keep silent. In uniform MAC, the resource allocation function $r(n) = \frac{1}{n}$, and thus the utility function:

$$U_{j\{uni\}}^{i} = \frac{\Psi_i}{n_i}. \qquad (3.23)$$

Note that $f_{uni}(n) = 1$. It is shown in [14] that such a game using uniform MAC does have the pure NE. In Proposition 1, we obtain the condition of the pure NE when uniform MAC is employed, and show that there may exist multiple NE-*sets*.

Proposition 1. *For the spectrum access game Γ using uniform MAC, if a congestion vector $\mathbf{n} = (n_1, n_2, \ldots, n_C)$ yields NE-set($\mathbf{n}$), the following condition should be satisfied:*

$$\begin{cases} n_i = \lceil \frac{\Psi_i N - \sum_{k \neq i, k \in \mathscr{C}} \Psi_k}{\sum_{k \in \mathscr{C}} \Psi_k} \rceil + W_0, \ i = 1, 2, \ldots, C \\ \sum_{i=1}^{C} n_i = N, \end{cases} \qquad (3.24)$$

where $W_0 \in \{0, 1, 2, \ldots, \lceil \frac{\Psi_i |N| + \Psi_i (|\mathscr{C}| - 1)}{\sum_{k \in \mathscr{C}} \Psi_k} \rceil - \lceil \frac{\Psi_i |N| - \sum_{k \neq i, k \in \mathscr{C}} \Psi_k}{\sum_{k \in \mathscr{C}} \Psi_k} \rceil - 1\}$. See the proof in section "NE Condition for Uniform MAC". From (3.24), it can be seen that there may exist more than one NE-*set*.

3.3.4 Slotted ALOHA

Compared with uniform MAC, slotted ALOHA is a more typical MAC used in ad hoc networks, including VANETs. In slotted ALOHA, users access the channel with probability p, and the throughput of each user is $th(p) = p(1-p)^{n-1}$. To maximize the throughput, let $th'(p) = 0$. Then, we get $p = \frac{1}{n}$, and the resource allocation function using slotted ALOHA is:

$$r_{SA}(n) = \frac{1}{n}(1 - \frac{1}{n})^{n-1}. \tag{3.25}$$

It can be shown that for slotted ALOHA, $f_{SA}(n) = (1 - \frac{1}{n})^{n-1}$, with $f'_{SA}(n) < 0$ and $f''_{SA}(n) > 0$. (See the proof in [14].) Moreover, if n goes to infinity, the total throughput of slotted ALOHA:

$$\lim_{n \to \infty} f_{SA}(n) = \frac{1}{e}. \tag{3.26}$$

The utility of VU j choosing channel i using slotted ALOHA is given by:

$$U^i_{j\{SA\}} = \Psi_i \frac{1}{n_i}(1 - \frac{1}{n_i})^{n_i-1}. \tag{3.27}$$

Different from uniform MAC, it is more difficult to derive the explicit condition of pure NE using slotted ALOHA. However, we show the existence of the pure NE and propose a scheme to achieve it.

Proposition 2. *In the spectrum access game with VU set $\mathcal{N}$ and channel set $\mathcal{C}$, each VU sequentially chooses the access channel. In each round, one VU chooses the best response to the strategies of the vehicles before him/her as the channel to access, i.e., its strategy in this game. Then, in each round, the strategy profile of the VU who have already made the decision is a pure NE.*

The proof is given in section "Proof of Proposition 2". Proposition 2 shows the existence of the pure NE in the spectrum access game Γ when using slotted ALOHA and provides a simple way to achieve a pure NE. However, to better understand the utilization of the channel resource, the efficiency of different NEs should be analyzed.

3.3.5 Efficiency Analysis

In the previous subsection, we prove the existence of pure NE(s) in the spectrum access game Γ. Generally speaking, an NE does not achieve global optimality due to the selfish behaviors of the players. In this subsection, we analyze the efficiency of an NE to evaluate the utilization of resources, which is defined as the total utility of all players under this NE. According to (3.22), in the spectrum access game Γ, the efficiency of a pure NE is defined as:

$$\mathcal{E}_S = \sum_{j=1}^{N} U_j^i = \sum_{j=1}^{N} \Psi_{s_j} r(n_{s_j}) = \sum_{i=1}^{C} \Psi_i n_i r(n_i), \tag{3.28}$$

where S is a strategy profile that constitutes a pure NE.

The social optimality is defined as the maximum total utility of all player among all possible strategy profiles. For a specific game, the social optimality is fixed. It is proved in [14] that the social optimality in Γ is:

$$opt_\Gamma = \begin{cases} \sum_{i=1}^{N} \Psi_i, & \text{if } N \leq C; \\ \sum_{i=1}^{C-1} \Psi_i + \sum_{i=C}^{N} \Psi_C r(N-C+1), & \text{if } N > C, \end{cases} \tag{3.29}$$

where Ψ_i is ordered such that $\Psi_1 \geq \Psi_2 \geq \cdots \geq \Psi_C$. Thus, to evaluate the efficiency of an NE, we define efficiency ratio (ER) of an NE as the ratio between the efficiency and the social optimality:

$$ER_S = \frac{\mathcal{E}_S}{opt_\Gamma}. \tag{3.30}$$

Different NE-*sets* may achieve different ERs. For example, in a game using uniform MAC with two channels ($\Psi_1 = 30$ and $\Psi_2 = 10$) and three vehicles, there are two NE-*sets*, as shown in Table 3.2, as well as their efficiency ratios. Obviously, NE-*set*$_2$ can achieve a higher ER than NE-*set*$_1$. In the following, the ER of the pure NE using uniform MAC and slotted ALOHA is discussed, respectively.

3.3.5.1 Uniform MAC

In uniform MAC, $f(n) = 1$. Among multiple NE-*sets* with different congestion vectors, we can easily draw the following conclusions:

- A pure NE in which each channel is chosen by at least one vehicle has ER=1.
- For any two different NE-*sets* NE-*set*$_1$ and NE-*set*$_2$ in which not all channels are chosen, if

Table 3.2 Multiple NE-*sets* in A game

	n1	n2	ER
NE-*set*$_1$	3	0	0.75
NE-*set*$_2$	2	1	1

$$\sum_{i=1}^{C} \Psi_i I_i^1 \geq \sum_{i=1}^{C} \Psi_i I_i^2, \tag{3.31}$$

where I_i^j is the indicator of whether channel i is chosen in NE-set_j, then $ER_1 \geq ER_2$.

The proof is straightforward. When uniform MAC is employed, the efficiency equals the summation of the ESA of all channels that are selected, i.e., $\mathcal{E}_S = \sum_{i=1}^{C} \Psi_i I_i^S$. When all channels are selected, all resources are fully utilized, and therefore, ER $= 1$. Otherwise, the higher efficiency yields higher efficiency ratio since the social optimality is fixed for a specific game.

3.3.5.2 Slotted ALOHA

In slotted ALOHA, although there is no explicit relation between the congestion vector and ER, Corollary 3.3.1 can help to lead a pure NE with high ER.

Corollary 3.3.1. *When Slotted ALOHA is used, in the process of composing a pure NE described in Proposition 2, the following rules can yield an NE with the highest efficiency ratio.*
If in a round the new vehicle has two best responses (BE_1 and BE_2),

- *when BE_1 corresponds to a vacant channel (no vehicle chooses it) and BE_2 corresponds to a channel that has been already chosen, then BE_1 is preferred;*
- *when each channel has been selected by at least one vehicle, the channel with **higher** ECA is preferred.*

See the proof in section "Proof of Corollary 3.3.1".

3.3.6 Distributed Algorithms to Achieve NE with High ER

After spectrum sensing, each VU has the knowledge of the available channels $i \in \mathcal{C}$, and the ECA of each channel, i.e., Ψ_i. VUs maintain a sorted list of the channels in $\mathcal{C}$ in a decreasing order of Ψ. Then, they participate in the distributed spectrum access game Γ. Since VUs behave in a distributed manner in CR-VANETs, the best solution to the game is a pure NE in which each VU has no incentive to change his/her current choice of the access channel unilaterally. According to Proposition 2, the pure NE can be achieved by each VU choosing the best response sequentially. Moreover, based on the analysis of Sect. 3.3.5 and Corollary 3.3.1, a pure NE with high ER can be achieved.

However, in such a process to achieve the NE, the VUs who choose their strategy before others usually benefit more. For instance, in a game Γ with two channels and two VUs, and $\Psi_1 = 15$ and $\Psi_2 = 10$, the one making decision first could obtain utility of 15 while the other VU could only obtain 10. To solve the problem, and achieve a pure NE with a high ER in a distributed manner, we design a distributed cognitive spectrum access algorithm, as shown in Algorithm 1. Each VU randomly selects a back-off time and starts the back-off. When the back-off timer expires, a VU chooses one channel to access according to the best response to the strategies of VUs that have already chosen the channel. Then, the VU broadcasts its decision in order for other VUs to derive their strategies. Since the selection of the back-off time is random, the proposed algorithm is fair for each VU.

Algorithm 1 Distributed Cognitive Spectrum Access Algorithm

1: // **Initialization**
2: Get available channels $\mathscr{C}$ by sensing.
3: Update and order the channel availability $[\Psi_1, \Psi_2, \ldots, \Psi_C]$ decreasingly using (3.17) and (3.18). Consider the current time is t_s.
4: Each vehicle that seeks for transmission opportunity picks a random back-off time t_b from $(0, t_{b\_max}]$, and starts the back-off.
5: **while** current time$\leq (t_s + t_{b\_max})$ **do**
6:     **if** The back-off timer of vehicle i expires **then**
7:         **if** uniform MAC **then**
8:             Select the best response with free channel considering the strategies that it receives. If all channels have been already chosen, then select any best response.
9:         **end if**
10:        **if** Slotted ALOHA **then**
11:            Select the channel according to Corollary 3.3.1.
12:        **end if**
13:        Broadcast the channel sequence number that it chooses.
14:    **end if**
15: **end while**
16: Each vehicle tunes its radio to its strategic channel, and starts transmission using specific MAC.
17: **return**

3.4 Performance Evaluation

In this section, the performance of the proposed congestion game based opportunistic spectrum access scheme is evaluated. We consider an urban scenario with $10\,\text{km} \times 10\,\text{km}$, where PTs and VUs (SUs) coexist. There are five licensed channels each with bandwidth of $1\,\text{MHz}$ which VUs can access in an opportunistic manner. PTs operating on different channels are associated with different parameters, i.e., R, L_P, λ_{idle}, and λ_{busy}. The length of road segment L is set to $100\,\text{m}$. Vehicles move in the area with a constant speed $v \in [10, 30]\,\text{m/s}$. The probabilities of vehicles selecting a direction at the intersection are set to $P_n = P_s = P_e = P_w = 0.25$.

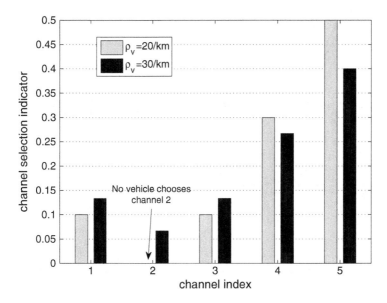

Fig. 3.9 Impact of ρ_v on Nash equilibrium

Denote by Th_j the utility of VU j by accessing the selected channel before the channel becomes unavailable, and the fairness index is defined as:

$$F = \frac{(\sum_j Th_j)^2}{N \sum_j Th_j^2},\tag{3.32}$$

which is used to evaluate the fairness among VUs [15]. Specifically, we compare the proposed spectrum access (denoted by 'NE') with a random channel access (denoted by 'random') in which VUs uniformly choose a channel from $\mathscr{C}$ to access.

Figure 3.9 shows the impact of vehicle density on the road (ρ_v) on the NE of the game, when uniform MAC is used. ρ_v captures the average number of vehicles on the road in unit length. Define the channel selection indicator of channel i as the ratio between the number of VUs choosing channel i and the total number of VUs, i.e., n_i/N, which indicates the popularity of the channel. When the value of ρ_v is small, some channels may not be chosen by any VU (e.g., channel 2 in Fig. 3.9 when ρ_v is 20/km). When the density of vehicles increases, all channels are selected by at least one VU and the selection indicator of each channel also changes to satisfy the NE condition (Proposition 1).

Figure 3.10 shows the performance of the proposed spectrum access scheme with respect to the vehicle speed v, when ρ is set to 20/km. From Fig. 3.10, it can be seen that 'NE' outperforms 'random' on average utility for both uniform MAC and slotted ALOHA. This is because for uniform MAC, in 'NE', VUs access the channels based on a pure NE while in 'random', each VUs chooses a channel in

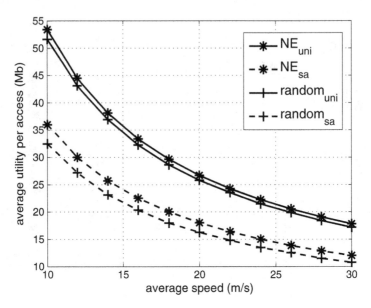

Fig. 3.10 Performance w.r.t. speed. $\rho_v = 20$/km

random manner, which may result in lower average utility since there may exist some channels which are not selected by any VU. However, for slotted ALOHA, channels with larger ECA are chosen by more VUs, resulting in more collisions. When vehicle density is 20/km, the decrease of resource utilization caused by collisions is less than that caused by random access in which some channels are not utilized, which accounts for the fact that 'NE' outperforms 'random' on average utility when using slotted ALOHA. The utilities of both 'NE' and 'random' decrease with the increase of vehicle speed, because a higher speed leads to a smaller channel availability Ψ, and thus a smaller average utility.

Figure 3.11 shows the performance of the proposed spectrum access scheme in terms of the vehicle density. From Fig. 3.11a, it can be seen that the average utility decreases with the vehicle density. This is straightforward since the total channel resource is fixed and the resource allocation function $r(n)$ is decreasing with n. For uniform MAC, the reason that 'NE' achieves higher average utility than 'random' is that in 'NE', vehicles always choose channels with higher ECA while in 'random' vehicles randomly choose the access channel. When ρ_v increases, the probability that all channels are chosen by at least one channel increases. Note that if each channel is selected by at least one VU, the average utility of 'NE' and 'random' is the same. This explains the reason that the difference of average utility between 'NE$_{uni}$' and 'random$_{uni}$' becomes smaller when vehicle density increases. For slotted ALOHA, when ρ_v increases, the average utility of 'NE' and 'random' also becomes closer, and the average utility of 'random' is slightly higher than that of 'NE' ($\rho_v >$ 20/km) because in 'NE', the preference to choose channels with higher ECA results in more collisions.

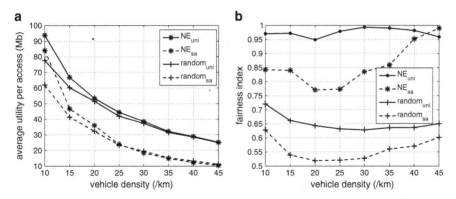

Fig. 3.11 Performance w.r.t. vehicle density. Vehicle speed $v = 10$ m/s. (**a**) Average utility w.r.t. vehicle density; (**b**) fairness w.r.t. vehicle density

Figure 3.11b shows the fairness index of 'NE' and 'random' in terms of vehicle density. It is shown that NE outperforms random access in terms of fairness in both uniform MAC and slotted ALOHA. This is because in a pure NE, the selfish property of VUs leads to an even share of the spectrum resources. On the other hand, in random access, each VU randomly chooses a channel to access, which results in different utilities among them. For 'NE$_{sa}$', when ρ_v is low, the increase of ρ_v may make VUs choose channels with different ECA to achieve NE, resulting in the decrease of fairness. For example, two VUs may both choose the channel with largest ECA, and the fairness index is 1. When ρ_v increases, a third VU may choose another channel, which makes the fairness index decrease. However, when ρ_v is high, with all channels selected, the increase of the number of VUs will make utilities among VUs closer based on the NE condition. If the density is extremely high, from (3.26), the game using slotted ALOHA turns to be a game using uniform MAC, with the channel bandwidth $\frac{1}{e}$ of the original bandwidth.

Figure 3.12 shows the efficiency ratio of the obtained NE in the proposed spectrum access game Γ. We introduce the channel diversity, which is defined as

$$\Phi = \sum_{i=1}^{C} (\Psi_i - \bar{\Psi})^2. \tag{3.33}$$

where $\bar{\Psi}$ is the mean ECA of all channels. Channel diversity indicates the variance among primary channels due to the properties of PTs, such as the spatial features and temporal channel usage pattern. A smaller value of Φ indicates that on average, the channels have relatively similar ECAs, and vice versa. Figure 3.12a shows the ER with respect to vehicle density. It can be seen that 'NE' achieves a higher ER than 'random' by utilizing either uniform MAC or slotted ALOHA, because the total utility of 'NE' is higher, as shown in Fig. 3.11a. The decrease of ER using random access when ρ_v is low is because with the increase of ρ_v, the social optimality also increases. However, for 'random', more VUs do not lead to as much increase in total

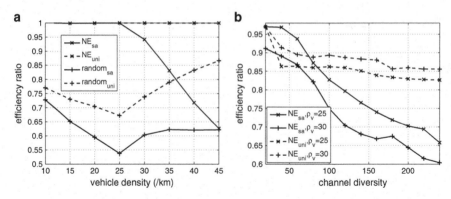

Fig. 3.12 Efficiency ratio. $\bar{\Psi} = 20$. (**a**) Efficiency ratio w.r.t. vehicle density. $\Omega = 20$; (**b**) efficiency ratio w.r.t. channel diversity

utility as in social optimality. The reason for the increase of the ER of 'random$_{uni}$' when vehicle density increases ($\rho_v \geq 25/km$) is that the social optimality will not change (all channels are selected) with the increase of ρ_v, while the total utility increases due to that in expectation, more channels are chosen. In fact, it approaches to the ER of 'NE$_{uni}$', which is not shown in the figure. The ER of 'random$_{sa}$' changes slightly when vehicle density is high. This is because when ρ_v is high, $f_{SA}(n)$ changes very little with ρ_v, and thus the total utility changes little due to (3.22). And with (3.26), when vehicle density is extremely high, we have

$$ER_{SA} \rightarrow \frac{\frac{1}{e}\sum_{i=1}^{C}\Psi_i}{\sum_{i=1}^{C-1}\Psi_i + \frac{1}{e}\Psi_C}, \tag{3.34}$$

where (3.34) is the lower bound of ER_{SA} when vehicle density increases.

Figure 3.12b shows the relation between ER and the channel diversity Φ. The ER of uniform MAC remains stable when Φ increases, because although the channels with smaller ECA are chosen less often, they have little impact on the ER since their ECA are small. However, for slotted ALOHA, the reason for the decrease of ER is twofold: first, the channels with smaller ECA are rarely chosen; second, more VUs choose channels with higher ECA, which leads to more contentions and collisions. When Φ increases, more VUs contend for the channels with higher ECA, resulting in more collisions, and smaller value of ER.

3.5 Summary

In this chapter, we have analyzed the channel availability for VUs in urban CR-VANETs, jointly considering the mobility of vehicles, and the spatial distribution and temporal channel usage pattern of PTs. We have then proposed a game theoretic spectrum access scheme to better exploit the spatial and temporal spectrum resource. Simulation results have demonstrated that the proposed scheme can achieve higher average utility and fairness than the random access scheme, when applying either uniform MAC or slotted ALOHA. The impact of vehicle density and channel diversity on the performance of the proposed scheme has also been studied. The research results can be applied for designing efficient spectrum sensing and access schemes in CR-VANETs.

Appendix

NE Condition for Uniform MAC

First, the situation for two channels is considered. Since $C = 2$ and $N \in \mathbb{Z}^+$, we have

$$\frac{\Psi_1}{n_1} \geq \frac{\Psi_2}{n_2 + 1} \text{ and } \frac{\Psi_2}{n_2} \geq \frac{\Psi_1}{n_1 + 1}, \tag{3.35}$$

which can be rewritten as follows:

$$\frac{\Psi_1}{\Psi_2} n_2 - 1 \leq n_1 \leq \frac{\Psi_1}{\Psi_2} n_2 + \frac{\Psi_1}{\Psi_2}. \tag{3.36}$$

Substitute $n_2 = N - n_1$ into (3.36), we obtain

$$\frac{\Psi_1 N - \Psi_2}{\Psi_1 + \Psi_2} \leq n_1 \leq \frac{\Psi_1 N + \Psi_1}{\Psi_1 + \Psi_2}. \tag{3.37}$$

Since

$$\frac{\Psi_1 N + \Psi_1}{\Psi_1 + \Psi_2} - \frac{\Psi_1 N - \Psi_2}{\Psi_1 + \Psi_2} = 1 \tag{3.38}$$

and

$$-1 < \frac{\Psi_1 N - \Psi_2}{\Psi_1 + \Psi_2} < N \tag{3.39}$$

Γ has at least one pure NE, in which

$$n_1 = \left\lceil \frac{\Psi_1 N - \Psi_2}{\Psi_1 + \Psi_2} \right\rceil \text{ and } n_2 = N - n_1. \tag{3.40}$$

Next, we extend this conclusion to the situation where more than two channels are available, i.e., $C > 2$. When $C > 2$, any two arbitrary channels i and k, $i, k \in \mathscr{C}$ should satisfy (3.36) to constitute an NE. Thus,

$$\frac{\Psi_k}{\Psi_i}n_i - 1 \leq n_k \leq \frac{\Psi_k}{\Psi_i}n_i + \frac{\Psi_k}{\Psi_i}. \tag{3.41}$$

Define $F_{L,ki}$ and $F_{U,ki}$ as

$$F_{L,ki} = \frac{\Psi_k}{\Psi_i}n_i - 1 \text{ and } F_{U,ki} = \frac{\Psi_k}{\Psi_i}n_i + \frac{\Psi_k}{\Psi_i}. \tag{3.42}$$

Then, for channels i and $\forall k \neq i, i, k \in \mathscr{C}$, we have

$$F_{L,ki} \leq n_k \leq F_{U,ki}. \tag{3.43}$$

It holds that

$$\sum_{k \neq i, k \in \mathscr{C}} F_{L,ki} \leq \sum_{k \neq i, k \in \mathscr{C}} n_k \leq \sum_{k \neq i, k \in \mathscr{C}} F_{U,ki}. \tag{3.44}$$

By substituting $\sum_{k \neq i, k \in \mathscr{C}} n_k = N - n_i$ into (3.44), we have

$$\frac{\Psi_i N - \sum_{k \neq i, k \in \mathscr{C}} \Psi_k}{\sum_{k \in \mathscr{C}} \Psi_k} \leq n_i \leq \frac{\Psi_i N + \Psi_i(C - 1)}{\sum_{k \in \mathscr{C}} \Psi_k}. \tag{3.45}$$

Similar to (3.38) and (3.39), it can be proved that

$$\frac{\Psi_i N + \Psi_i(C - 1)}{\sum_{k \in \mathscr{C}} \Psi_k} - \frac{\Psi_i N - \sum_{k \neq i, k \in \mathscr{C}} \Psi_k}{\sum_{k \in \mathscr{C}} \Psi_k} > 1 \tag{3.46}$$

and

$$-1 < \frac{\Psi_i N - \sum_{k \neq i, k \in \mathscr{C}} \Psi_k}{\sum_{k \in \mathscr{C}} \Psi_k} < N. \tag{3.47}$$

Then, for any $\mathscr{C}$ and $\mathscr{N}$, (3.18) has at least one solution, which is

$$n_i = \lceil \frac{\Psi_i N - \sum_{k \neq i, k \in \mathscr{C}} \Psi_k}{\sum_{k \in \mathscr{C}} \Psi_k} \rceil + W_0, \tag{3.48}$$

where $W_0 \in \{0, 1, 2, \ldots, \lceil \frac{\Psi_i N + \Psi_i(C-1)}{\sum_{k \in \mathscr{C}} \Psi_k} \rceil - \lceil \frac{\Psi_i N - \sum_{k \neq i, k \in \mathscr{C}} \Psi_k}{\sum_{k \in \mathscr{C}} \Psi_k} \rceil - 1\}$. With $\sum_{i \in \mathscr{C}} n_i = N$, we have (3.24). Thus, the game Γ has at least one pure NE (3.24) is called NE condition of the spectrum access game Γ when uniform MAC is used.

Proof of Proposition 2

Assume that for a given round R_t, the congestion vector $\mathbf{n}(S_t) = \{n_1, n_2, \ldots, n_C\}$ composes a pure NE. According to (3.18), for each channel $i \in \mathscr{C}$,

$$\Psi_i r(n_i) \geq \Psi_k r(n_k + 1), \ \forall k \in \mathscr{C}, k \neq i. \tag{3.49}$$

Then for a new round R_{t+1}, a new vehicle joins the game and chooses its best response according to the existing strategy profile, i.e., $\mathbf{n}(S_t)$. Consider its best response is channel m, and thus the new congestion vector is $\mathbf{n}(S_{t+1}) = \{n_1, \ldots, n_m + 1, \ldots, n_C\}$. For the new congestion vector, we have the following observations:

1. For each channel $i \in \mathscr{C}, i \neq m$, $\Psi_i r(n_i) \geq \Psi_k r(n_k + 1), \ \forall k \in \mathscr{C} \setminus \{i, m\}$ holds because the number of vehicles that choose the channels other than channel m does not change, and $r(n_i), i \neq m$ remains unchanged.
2. $\Psi_m r(n_m + 1) \geq \Psi_k r(n_k + 1), \forall k \in \mathscr{C}, k \neq m$. This statement holds due to that channel m is the best response for the new vehicle.
3. $\Psi_k r(n_k) \geq \Psi_m r(n_m + 1 + 1), \forall k \in \mathscr{C}, k \neq m$. Remember in round t, $\Psi_k r(n_k) \geq \Psi_m r(n_m + 1)$. $r(n)$ is a non-increasing function, and thus $r(n_m + 1) \geq r(n_m + 1 + 1)$.

Therefore, $\mathbf{n}(S_{t+1})$ also constitutes a pure NE. For a specific game, the first vehicle chooses the channel with largest ECA and of course composes a pure NE. Then, for each round, the strategies of vehicles which have participated in the game constitute a new pure NE, until all vehicles have chosen their strategies.

Proof of Corollary 3.3.1

For any round in Proposition 2, assume that the congestion vector $\mathbf{n}(S) = \{n_1, n_2, \ldots, n_C\}$ constitutes a pure NE and Ψ_i is sorted so that $\Psi_1 \geq \Psi_2 \geq \cdots \geq \Psi_C$. The efficiency of the NE is

$$\mathscr{E}_S = \sum_{i=1}^{C} f(n_i). \tag{3.50}$$

Remember that in slotted ALOHA, $f(n) = (1 - \frac{1}{n})^{n-1}$. A new vehicle comes and finds there are more than one best response (BR).

1. If BR_1 corresponds to a free channel i when BR_2 corresponds to channel j that has been selected by at least one vehicle, then BR_1 leads to a NE with efficiency:

$$\mathscr{E}_{S1} = \mathscr{E}_S + \Psi_i > \mathscr{E}_S. \tag{3.51}$$

BR_2 leads to a NE with efficiency:

$$\mathscr{E}_{S2} = \mathscr{E}_S - \Delta < \mathscr{E}_S. \tag{3.52}$$

where Δ is the loss of $f(n_j)$ since $f(n)$ decreases with n. Obviously, $\mathscr{E}_{S1} > \mathscr{E}_{S2}$.

2. Consider that BR_1 and BR_2 correspond to channel i and j with $n_i \geq 1$ and $n_j \geq 1$, respectively. Without loss of generality, consider $\Psi_i > \Psi_j$. Under this condition, it is clear that $n_i > n_j$, or else channel i and j cannot be the best response simultaneously. Consider only the total utility of users choosing channel i and j since other channels are not affected in this round. BR_1 will lead to an NE with utility $\mathscr{E}_1 = \Psi_i f(n_i + 1) + \Psi_j f(n_j)$, while BE_2 will lead to an NE with utility $\mathscr{E}_2 = \Psi_i f(n_i) + \Psi_j f(n_j + 1)$. Using the property of the pure NE, we have $\Psi_i r(n_i + 1) \geq \Psi_j r(n_j + 1)$ and $\Psi_j r(n_j + 1) \geq \Psi_i r(n_i + 1)$, and thus $\Psi_i r(n_i + 1) = \Psi_j r(n_j + 1)$, i.e., $\Psi_i \frac{f(n_i+1)}{n_i+1} = \Psi_j \frac{f(n_j+1)}{n_j+1}$. Let

$$\Psi_i = \frac{\frac{f(n_j+1)}{n_j+1}}{\frac{f(n_i+1)}{n_i+1}} \Psi_j = \alpha \Psi_j. \tag{3.53}$$

To prove

$$\begin{aligned} \mathscr{E}_1 - \mathscr{E}_2 &= \Psi_i f(n_i + 1) + \Psi_j f(n_j) - (\Psi_i f(n_i) + \Psi_j f(n_j + 1)) \\ &= \Psi_j[\alpha(f(n_i + 1) - f(n_i)) + f(n_j) - f(n_j + 1)] > 0, \end{aligned} \tag{3.54}$$

is equivalent to prove

$$\alpha = \frac{\frac{f(n_j+1)}{n_j+1}}{\frac{f(n_i+1)}{n_i+1}} < \frac{f(n_j) - f(n_j + 1)}{f(n_i) - f(n_i + 1)}, \tag{3.55}$$

since $f(n) - f(n + 1) > 0$.

$$\frac{\frac{f(n_j+1)}{n_j+1}}{\frac{f(n_i+1)}{n_i+1}} < \frac{f(n_j) - f(n_j + 1)}{f(n_i) - f(n_i + 1)}$$

$$\Leftrightarrow \frac{\frac{f(n_j+1)}{n_j+1}}{f(n_j) - f(n_j + 1)} < \frac{\frac{f(n_i+1)}{n_i+1}}{f(n_i) - f(n_i + 1)}$$

$$\Leftrightarrow g(n) = \frac{\frac{f(n+1)}{n+1}}{f(n) - f(n + 1)} \text{ increasing with } n \geq 1$$

$$\Leftrightarrow g'(n) > 0, \text{ when } n \geq 1. \tag{3.56}$$

We skip the tedious proof of (3.56) to simplify the exposition. Then, we have $\mathscr{E}_1 > \mathscr{E}_2$.

References

1. Haykin S (2005) Cognitive radio: brain-empowered wireless communications. IEEE J Sel Areas Commun 23(2):201–220
2. Zhang N, Lu N, Lu R, Mark J, Shen X (2012) Energy-efficient and trust-aware cooperation in cognitive radio networks. In: Proceedings of IEEE ICC, Ottawa, June 2012
3. Cheng N, Zhang N, Lu N, Shen X, Mark J, Liu F (2014) Opportunistic spectrum access for cr-vanets: a game-theoretic approach. IEEE Trans Veh Technol 63(1):237–251
4. Kostof S, Tobias R (1991) The city shaped. Thames and Hudson, London
5. Siksna A (1997) The effects of block size and form in north american and australian city centres. Urban Morphol 1:19–33
6. Lee J, Mazumdar R, Shroff N (2006) Joint resource allocation and base-station assignment for the downlink in CDMA networks. IEEE/ACM Trans Netw 14(1):1–14
7. Liu Y, Cai L, Shen X (2012) Spectrum-aware opportunistic routing in multi-hop cognitive radio networks. IEEE J Sel Areas Commun 30(10):1958–1968
8. Chun Ting C, Sai S, Hyoil K, Kang G (2007) What and how much to gain by spectrum agility? IEEE J Sel Areas Commun 25(3):576–588
9. Min A, Kim K, Singh J, Shin K (2011) Opportunistic spectrum access for mobile cognitive radios. In: Proceedings of IEEE INFOCOM, Shanghai, April 2011
10. Niyato D, Hossain E, Wang P (2011) Optimal channel access management with qos support for cognitive vehicular networks. IEEE Trans Mob Comput 10(5):573–591
11. Zhang X, Su H, Chen H (2006) Cluster-based multi-channel communications protocols in vehicle ad hoc networks. IEEE Wirel Commun 13(5):44–51
12. Han Z, Niyato D, Saad W, Başar T, Hjørungnes A (2011) Game theory in wireless and communication networks: theory, models, and applications. Cambridge University Press, New York
13. Blumrosen L, Dobzinski S (2007) Welfare maximization in congestion games. IEEE J Sel Area Commun 25(6):1224–1236
14. Laq LM, Huang J, Liu M (2012) Price of Anarchy for Congestion Games in Cognitive Radio Networks. IEEE Trans Wireless Commun 11(10):3778–3787
15. Jain R, Chiu D, Hawe W (1984) A quantitative measure of fairness and discrimination for resource allocation in shared computer systems. DEC research report TR-301

Chapter 4
Performance Analysis of WiFi Offloading in Vehicular Environments

The demand for high-speed mobile Internet services has dramatically increased. In a recent survey [1], it is revealed that Internet access is predicted to become a standard feature of future vehicles. Cellular-based access technologies, such as LTE, play a critical role in providing ubiquitous and reliable Internet access to vehicles, as the cellular infrastructure is well planned and widely available. However, as we mentioned in Chap. 1, the cellular network nowadays is already congested due to the explosive mobile data demand. Therefore, simply employing cellular networks for vehicular Internet access may worsen the congestion, and consequently degrade the performance of both non-vehicular and vehicular users.

With millions of hotspots deployed around the world, WiFi can be a complementary and cost-effective solution to vehicular Internet access. The existing research work on drive-thru Internet and vehicular WiFi offloading has been surveyed in Chap. 2. In addition, recent advances in Passpoint/Hotspot 2.0 make WiFi more suitable to provide secure connectivity and seamless roaming [2]. In a vehicular environment, vehicles can signal to nearby WiFi access points (APs) when moving along a road, such that the cellular traffic can be delivered to vehicles through the drive-thru Internet in an opportunistic manner. Opportunistic vehicular WiFi offloading has unique features:

- In each drive-thru, a relatively small volume of data can be delivered from/to a vehicle, due to the short connection time with WiFi APs; and
- The offloading performance can be significantly improved if the data services/applications can tolerate a certain delay, as vehicles with a high speed can have multiple drive-thru opportunities in a short future.

To the best of our knowledge, the WiFi offloading performance in the vehicular environment remains unclear in the literature. In this chapter, we aim to theoretically analyze the performance of vehicular WiFi offloading with the delayed offloading strategy. We consider a generic VU having randomly arriving data services, either to

© The Author(s) 2016
N. Cheng, X. (Sherman) Shen, *Opportunistic Spectrum Utilization in Vehicular Communication Networks*, SpringerBriefs in Electrical and Computer Engineering, DOI 10.1007/978-3-319-20445-1_4

download data (e.g., E-mail attachment and YouTube video clip) from the Internet, or upload data (e.g., WeChat messages and online diagnosis data) to the Internet. The data services can be fulfilled via either the cost-effective WiFi networks (want-to) or the cellular network (have-to). The cellular network is considered to provide full coverage and WiFi APs are sparsely deployed. The offloading performance is characterized by two metrics: average service delay and offloading effectiveness. *Average service delay* is defined as the average delay from the arrival of a data service request to the fulfillment of the service via WiFi networks, which is mostly contributed by the waiting time for the WiFi transmission opportunities and is the "price" of using delayed offloading. *Offloading effectiveness* is defined as the long-term proportion of data services (amount of data) fulfilled by WiFi networks under the requirement of a desired average service delay. Intuitively, if the data services can be deferred for a longer time (i.e., relaxing the requirement on the average service delay), more data services can be offloaded by WiFi networks, yielding a higher offloading effectiveness. From a VU's perspective, however, the increased average service delay would somehow reduce the user satisfaction on data services. The analytical framework proposed in this chapter provides the relation between these two metrics and examines the tradeoff.

In more detail, we develop a queueing analysis of the offloading performance in a vehicular environment. Each VU maintains an M/G/1/K queue, with M characterizing the Markovian arrival process of data service requests, and system capacity K controlling the average service delay. If a service request arrives and sees a full queue, it will be served directly via the cellular network to avoid an unacceptably long expected service delay. As the WiFi data transmission is in an opportunistic manner, the queue departure process (i.e., the process of data transmission or service fulfillment) is characterized by the *effective service time* (EST) which is the duration from the transmission of the first bit of a data service to the service request is fulfilled. The probability distribution of the EST is theoretically derived through Laplace-Stieltjes transform (LST). Based on the statistics of the EST, the offloading effectiveness and average service delay and their relation are given. Our analytical framework is validated through simulations based on a VANETs simulation tool VANETMobisim and real map data sets.

Implementation of our analytical framework: For VUs, the explicit relation between offloading effectiveness (how much Internet access cost can be saved) and average service delay (how much service degradation the user is willing to tolerate) can provide offloading guidelines and help to design offloading strategies. For example, the on-board smart offloading engine (a mobile App) can intelligently select access network based on the user's preference on service delay or cost requirements. On the other hand, due to the razor-sharp competition, many network operators are looking for new ways to cut spending and stand out in the market. Network operators, such as AT&T, NTT Docomo and China Mobile, are starting to deploy carrier-WiFi networks to offload cellular networks and profit from the new business model. Our results can give the network operators more incentives to deploy outdoor WiFi networks as the offloading effectiveness is notable in vehicular environments. Moreover, our framework can provide network operators with some guidance on WiFi deployment (e.g., the density of WiFi APs) according to the theoretical offloading effectiveness. In a nutshell, the analytical framework can be applied in practice not only for VUs to make offloading decisions, but also for network operators to evaluate AP deployment strategies, make offloading-related pricing models, and so forth.

The remainder of the chapter is organized as follows. Section 4.1 describes the system model. Section 4.2 derives the probability distribution of the EST. Section 4.3 analyzes the queue and the tradeoff between offloading effectiveness and average service delay. Section 4.4 evaluates the analysis by real road map based simulation. Section 4.5 summarizes the chapter.

4.1 System Model

In this section, we present the system model, including communication paradigm, mobility of vehicles, and queueing model of VUs. Based on the system model, we can evaluate the performance of vehicular WiFi offloading by analyzing the M/G/1/K queue.

4.1.1 Communication Model

We consider an urban area as a bounded region where WiFi APs are randomly deployed. VUs access drive-thru Internet when possible since it tends to have a higher data rate and less cost than cellular networks. Automatic rate adaptation is widely used in the stock WiFi technology according to the signal strength. However, for simplicity, we consider the communication data rate between APs and VUs is identical and denoted by R. Assume that the ideal MAC is employed where channel access time is fairly shared by the nearby VUs. To account for real MAC behaviors, a MAC throughput effective factor η is considered. η indicates the theoretical maximum portion of throughput considering the protocol overhead, e.g., $\eta = 45.5\%$ for bit rate 11 Mbps of IEEE 802.11b [3]. Thus, the data rate of a tagged VU $\mathcal{V}$ can be represented by $r = \frac{\eta R}{n+1}$, where n is the number of neighbor VUs of $\mathcal{V}$ connected to the same AP. With the second order Taylor approximation, the average data rate of an arbitrary VU connected to an AP can be approximated by

$$\bar{r} = r|_{\bar{n}} + \frac{1}{2}\mathrm{Var}(n)\frac{d^2 r}{dn^2}|_{\bar{n}}, \qquad (4.1)$$

where $\bar{n}$ and $\mathrm{Var}(n)$ are the mean and variance of n, respectively [4].

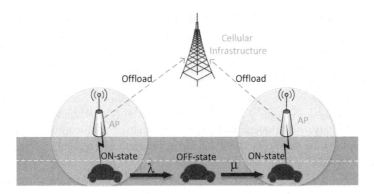

Fig. 4.1 System model

4.1.2 Mobility Model

The high mobility of vehicles may lead to short and intermittent drive-thru access opportunities. We model the mobility of $\mathcal{V}$ by an on-off process, with on-state and off-state denoting the situation that $\mathcal{V}$ is within and out of the coverage area of an AP. In on-state, $\mathcal{V}$ can transmit data through the AP with an average data rate $\bar{r}$; and in off-state, the transmission rate is assigned to zero, since $\mathcal{V}$ is out of the coverage area of any AP. Due to the random locations and coverage of open APs, we model the sojourn time of both on-state and off-state by the unpredictable and memoryless exponential distribution, with parameter λ and μ, respectively [4], as shown in Fig. 4.1.

4.1.3 Queueing Model

Each VU maintains an M/G/1/K queue to store the data service requests which are not served yet, as shown in Fig. 4.2. The arrival process of the queue is considered to follow Poisson process with mean inter-arrival time $1/\gamma$. We also consider that the sizes of service requests are identical, denoted by S. The departure process of the queue is not Markovian due to the service interruption. If a service request cannot be fulfilled within one drive-thru, it will wait for subsequent drive-thru opportunities until fulfillment. We model the departure process by the EST which follows a general distribution. The EST t_e composes of the service time x, and time of server interruptions, i.e., when the vehicle is out of the coverage area of any AP. The queue capacity K represents the maximum number of service requests in the system, and thus the queue buffer capacity is $K - 1$. We consider patient customer-type of queue, which means that if one service request enters the queue, it will wait until fulfillment. However, if a service arrives with a full queue, it expects that the delay would exceed the desired service delay. As a result, the service

Fig. 4.2 Queueing model for
data services of VUs

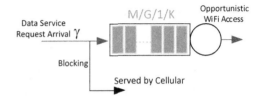

Data Service
Request Arrival γ

M/G/1/K

Opportunistic
WiFi Access

Blocking

Served by Cellular

request will be directly served by the cellular network. With the queueing model
and probability distribution of the EST which is obtained in the next section, the
queue performance can be evaluated, with the average service delay and offloading
effectiveness analyzed.

4.2   Derivation of Effective Service Time

We characterize the departure process and derive the probability distribution of the
EST in this section.

Since the data transmission cannot happen when a vehicle is outside the coverage
area of any WiFi AP, the queue server is subject to intermittent interruptions. Let
$t_e(x)$ denote the EST of an arbitrary request with service time x, whereas x is
assumed to follow the exponential distribution with parameter $\lambda_s = \frac{\bar{r}}{S}$, as in [4].
Since $t_e(x)$ composes of x and the time of server interruptions, we then have

$$t_e(x) = \begin{cases} x & H_\lambda \geq x \\ H_\lambda + H_\mu + T(x - H_\lambda) & H_\lambda < x, \end{cases} \tag{4.2}$$

where H_λ and H_μ are the length of an arbitrary on-state period and off-state period,
which follow the exponential distribution with mean $1/\lambda$ and $1/\mu$, respectively.
According to [5], the EST can be analyzed using LST. Let $T_e(\xi)$ be LST of the
probability density function (PDF) of $t_e(x)$, and based on [5]:

$$T_e(\xi) = \int_0^\infty f(x) e^{-(\xi + \lambda - \lambda V_\mu(\xi))} dx$$

$$= \int_0^\infty \lambda_s e^{-\lambda_s x} e^{-(\xi + \lambda - \lambda V_\mu(\xi))} dx$$

$$= \frac{\lambda_s}{\xi + \lambda - \lambda V_\mu(\xi) + \lambda_s}, \tag{4.3}$$

where $V_\mu(\xi) = \frac{\mu}{\mu + \xi}$ is LST of the PDF of the length of an arbitrary off-state period.
We can then obtain the expectation of the EST by

$$\mathbb{E}[t_e] = -\frac{dT_e(\xi)}{d\xi}\Big|_{\xi=0}$$

$$= \frac{\lambda_s(1 - \lambda\frac{dV_\mu(\xi)}{d\xi})}{[\xi + \lambda - \lambda V_\mu(\xi) + \lambda_s]^2}\Big|_{\xi=0}$$

$$= \frac{1}{\lambda_s}(1 + \frac{\lambda}{\mu}). \tag{4.4}$$

From (4.4), $\mathbb{E}[t_e]$ is affected by server interruptions in the way that $\mathbb{E}[t_e]$ increases with the increase of interruption occurrence rate (λ) and mean interruption duration ($\frac{1}{\mu}$).

To analyze the M/G/1/K queue, we then derive the PDF of the EST, which is denoted by $f_e(t)$. Utilizing the inverse transform of LST, we have

$$f_e(t) = \mathscr{L}^{-1}(T_e(\xi)) = \frac{1}{2\pi i}\lim_{\delta\to\infty}\int_{\gamma-i\delta}^{\gamma+i\delta} e^{\xi t}T_e(\xi)d\xi, \tag{4.5}$$

where θ is a real number that is greater than the real part of all singularities of $T_e(\xi)$. The singularities of $T_e(\xi)$ can be obtained by making the denominators equal zero, i.e.,

$$\xi_{sin} = \{\xi|\xi + \mu = 0\} \cup \{\xi|\xi + \lambda - \lambda\frac{\mu}{\mu + \xi} + \lambda_s = 0\}. \tag{4.6}$$

Simplifying (4.6), we can get $\xi_{sin1} = -\mu$ and $\xi_{sin2} = \frac{1}{2}[-(\lambda + \lambda_s + \mu) \pm \sqrt{(\lambda + \lambda_s + \mu)^2 - 4\lambda_s\mu}]$ (if $\xi + \lambda - \lambda\frac{\mu}{\mu+\xi} + \lambda_s = 0$ has solution(s).) Since all singularities of $T_e(\xi)$ are smaller than zero, we can set $\theta = 0$. Using Bromwich inversion integral [6] and the fact that $\theta = 0$, we have

$$f_e(t) = \frac{2e^{\theta t}}{\pi}\int_0^\infty Re(T_e(\theta + iu))\cos(ut)du$$

$$= \frac{2}{\pi}\int_0^\infty Re[\frac{\lambda_s(\mu + iu)}{iu(\mu + iu) + (\lambda + \lambda_s)(\mu + iu) - \lambda\mu}]\cos(ut)du$$

$$= \frac{2}{\pi}\int_0^\infty \frac{\lambda_s(\lambda_s\mu^2 + \lambda u^2 + \lambda_s u^2)}{(\lambda_s\mu - u^2)^2 + u^2(\lambda + \lambda_s + \mu)^2}\cos(ut)du. \tag{4.7}$$

Since integration (4.7) is difficult to calculate, we use numerical methods to calculate $f_e(t)$ for any given t_e. We use Fourier-Series method and the trapezoidal rule to obtain the numerical integration of (4.7) [6].

$$f_e(t) \approx f_e^h(t)$$

$$\equiv \frac{he^{\theta t}}{\pi}T_e(\theta) + \frac{2he^{\theta t}}{\pi}\sum_{k=1}^\infty Re(T_e(\theta + ikh))\cos(kht). \tag{4.8}$$

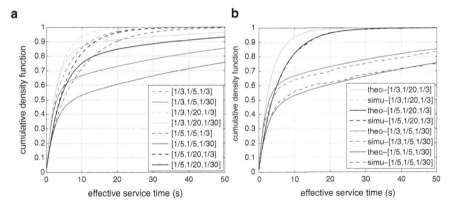

Fig. 4.3 Results of the EST. In the legend, [x,y,z] stand for λ_s, λ and μ, respectively. (**a**) EST w.r.t. λ_s, λ and μ; (**b**) theoretical and simulation results of EST

Some results of the EST derivation are shown in Fig. 4.3a, with the effects of λ_s, λ and μ compared. It can be seen that the EST increases with the decrease of λ_s and μ, and with the increase of λ. Smaller λ_s indicates a larger request service time x, and thus a larger EST with server interruptions. On the other hand, because a larger λ and a smaller μ indicate more frequent and longer interruptions, respectively, the EST then increases. To validate the accuracy of the EST derivation, the theoretical results are compared with the simulation results in Fig. 4.3b. In the simulation, a virtual queue with exhaustive customer arrivals and server interruptions is considered. From the figure, it can be shown that the curves of theoretical and simulation results closely match each other, which demonstrates the accuracy of the analysis.

4.3 Analysis of Queueing System and Offloading Performance

Given the probability distribution of the EST obtained in (4.8), we can evaluate the performance of offloading by analyzing the M/G/1/K queue. Because the EST is not exponential, we utilize the imbedded Markov chain of system states at the time instant t_{di}, which is the time instance of service request i's fulfillment, to analyze the queue [7].

4.3.1 Queue Analysis

Let n_i be the number of requests left in the system seen by the ith request when it leaves the system, and χ_i be the number of arrivals during the EST of request i. We then have

$$n_{i+1} = n_i - U(n_i) + \chi_{i+1}, \tag{4.9}$$

where $U(n_i)$ is the unit step function where $U(n_i) = 0$ if $n_i = 0$, and $U(n_i) = 1$ otherwise. Then, the transition probabilities at departure instances are defined as

$$p_{d,jk} = P\{n_{i+1} = k | n_i = j\} \quad 0 \le j, k \le K - 1. \tag{4.10}$$

Let ω_k be the probability of k arrivals during the EST of an arbitrary service request. Using the Poisson arrival property, we can get

$$\omega_k = \int_{t=0}^{\infty} \frac{(\gamma t)^k}{k!} e^{-\gamma t} f_e(t) dt, \tag{4.11}$$

where γ is the arrival rate of service requests and $f_e(t)$ is the PDF of the EST given in (4.8). Thus, we can easily obtain the $K \times K$ transition probability matrix as

$$\mathbf{P_{d,jk}} = \begin{bmatrix} \omega_0 & \omega_1 & \omega_2 & \cdots & \omega_{K-2} & \sum_{m=K-1}^{\infty} \omega_m \\ \omega_0 & \omega_1 & \omega_2 & \cdots & \omega_{K-2} & \sum_{m=K-1}^{\infty} \omega_m \\ 0 & \omega_0 & \omega_1 & \cdots & \omega_{K-3} & \sum_{m=K-2}^{\infty} \omega_m \\ \vdots & \vdots & \vdots & & \vdots & \vdots \\ 0 & 0 & 0 & \cdots & \omega_0 & 1 - \omega_0 \end{bmatrix} \tag{4.12}$$

Let $p_{d,k}$ be the equilibrium state probabilities at departure instants, which can be computed by

$$p_{d,k} = \sum_{j=0}^{K-1} p_{d,j} p_{d,jk} \quad k = 0, 1, \cdots, K - 1 \tag{4.13}$$

$$\sum_{k=0}^{K-1} p_{d,k} = 1 \quad (Normalisation \ Condition). \tag{4.14}$$

Let $p_k, k = 0, 1, 2, \cdots, K$ be the steady-state probabilities of the system states, and $P_B = p_K$ be the blocking probability. Based on Poisson arrivals see time averages (PASTA) property, p_k can be calculated by (4.15) [7].

$$p_k = (1 - P_B) p_{d,k} \quad k = 0, 1, 2, \cdots, K - 1. \tag{4.15}$$

The traffic intensity ρ and the actual traffic intensity ρ_c considering queue blocking are given by $\rho = \gamma \mathbb{E}[t_e]$ and $\rho_c = (1 - P_B)\rho$, respectively. It is straightforward that p_0, the steady-state probability that the queue is empty, should equal $1 - \rho_c$. Then, we can have

$$P_B = 1 - \frac{1}{p_{d,0} + \rho}, \tag{4.16}$$

by using (4.15) for the case $k = 0$ and the fact that $p_0 = 1 - \rho_c$. Using (4.15) and (4.16), we have

$$p_k = \frac{1}{p_{d,0} + \rho} p_{d,k} \quad k = 0, 1, 2, \cdots, K - 1. \tag{4.17}$$

4.3.2 Offloading Performance

Using (4.16) and (4.17), the mean number of customers N in the system can be represented as a function of K:

$$N(K) = \sum_{k=0}^{K} k p_k = \frac{1}{p_{d,0} + \rho} \sum_{k=0}^{K-1} k p_{d,k} + K(1 - \frac{1}{p_{d,0} + \rho}). \tag{4.18}$$

And the average total time in the system, i.e., average service delay, is also a function of K and can be calculated using Little's law:

$$W(K) = \frac{N}{(1 - P_B)\gamma} = \frac{\sum_{k=0}^{K-1} k p_{d,k} + K(p_{d,0} + \rho - 1)}{\gamma}. \tag{4.19}$$

We use unblocked rate $(1 - P_B)$ to measure the offloading effectiveness $\mathcal{E}$. For the system with queue capacity K, request size S, and statistics of on and off periods, λ and μ, the offloading effectiveness can be calculated by

$$\mathcal{E} = 1 - P_B = \frac{1}{p_{d,0} + \rho}. \tag{4.20}$$

Thus, the blocked services requests, with portion P_B of the total traffic, should be transmitted using cellular networks. To show the offloading capability for a given traffic load, define average offloading throughput for traffic load γ as

$$\Omega = \gamma S \mathcal{E} = \frac{\gamma S}{p_{d,0} + \rho}. \tag{4.21}$$

It can be seen that for a given γ, a larger $\mathcal{E}$ generally leads to a larger Ω. However, for an overload queue ($\rho > 1$) due to the heavy traffic, there is an upper bound of the average offloading throughput, denoted by Ω_m. For an overloaded queue, we can calculate Ω_m by setting $p_0 = 0$ since the server keeps busy serving the requests. Because the average service time of a request is its mean EST, calculated by (4.4), we can obtain the upper bound of the average offloading throughput as

$$\Omega_m = \frac{S}{\mathbb{E}[t_e]} = \frac{\bar{r}}{1 + \frac{\lambda}{\mu}}. \tag{4.22}$$

As discussed above, there is a tradeoff between average service delay and the offloading effectiveness. Such a tradeoff is analyzed using the results of the queueing model. There are two cases that VUs and network operators may care about: (1) Given a certain average service delay, how much data can be offloaded; (2) To offload a certain amount of data, how much is the least average delay that the users should tolerate. For (1), assume that the average service delay that users can tolerate is W_U^*. The corresponding K should be

$$K^* = \max\{K|W(K) \le W_U^*\}. \tag{4.23}$$

Then, the offloading effectiveness $\mathscr{E}^*$ can be calculated by

$$\mathscr{E}^* = 1 - P_B|_{K=K^*} = \frac{1}{p_{d,0} + \rho}|_{K=K^*}. \tag{4.24}$$

For (2), the solution is similar by setting a target offloading effectiveness $\mathscr{E}_U^*$.

4.4 Performance Evaluation

In this section, we evaluate the proposed queueing model, demonstrate the performance of vehicular WiFi offloading and examine the tradeoff between offloading effectiveness and average service delay. The simulation is carried out in a 2.0 km × 2.0 km region road map of the downtown area of Washington D.C., USA. WiFi APs are randomly deployed with the coverage radius of 100 m. There may exist overlapping of WiFi coverage areas. However, we show that low-level overlapping has little impact on the offloading performance. The street layout and AP locations are shown in Fig. 4.4. Each street segment has two lanes with the bidirectional vehicle traffic. VANETMobisim [9] is used to generate the mobility traces of 300 vehicles. With 50 deployed APs, the parameters can be obtained from the simulation, as $1/\lambda = 31.5$ s, $1/\mu = 52.09$ s, $\bar{n} = 1.54$, and $\mathrm{Var}(n) = 7.71$. It is considered that vehicles can transmit through WiFi immediately when they move into the coverage area, which is already supported by advanced WiFi technologies, e.g., HotSpot 2.0. With 802.11b bit rate 11 Mbps, $\bar{r} = 4.32$ Mbps using (4.1). Note that our proposed analytical model can be applied to other WiFi technologies if the average data rate is known. The service request size S is set to 5 MB, which is the size of a typical MP3 file.

The offloading effectiveness $\mathscr{E} = 1 - P_B$ is shown in Fig. 4.5a. It can be seen that with a larger queue capacity K, the offloading effectiveness $\mathscr{E}$ increases, which means that a larger portion of mobile data can be offloaded via WiFi networks. Such results are straightforward because with larger K, more requests can be temporarily buffered, and then served when WiFi is available. On the other hand, if K is smaller, more service requests are blocked due to a full queue. However, there is a tradeoff between the offloading effectiveness and the average service delay.

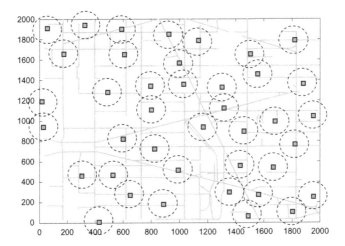

Fig. 4.4 Street layout and AP locations. Number of APs is 40. The map data is from TIGER/Line Shapefiles [8]

The average service delay is shown in Fig. 4.5b. It can be seen that a larger K leads to larger average service delay, which may decrease the user satisfaction. We can also see that a large value of request arrival rate γ increases the average service delay and decreases the offloading effectiveness due to a larger traffic load. Moreover, the curves of analysis and simulation results match each other well, which validates the theoretical analysis. Figure 4.5c shows the average offloading throughput Ω. It can be seen that Ω increases with K, since $\mathcal{E}$ increases. When the queue is overloaded (e.g., $\gamma = 0.11$ in Fig. 4.5c), the upper bound of the average offloading throughput in such a scenario is about 1.63 Mbps.

To better depict the tradeoff between service delay and offloading effectiveness, we introduce the concept of user satisfaction level Φ, which reflects the user satisfaction with respect to the average delay W. In this chapter, we simply use a linear relationship between Φ and W as $\Phi(W) = 1 - W/D_t$, where D_t is the delay tolerance of a user. Then, we define the utility of a VU as $H = \varpi\Omega + \Phi(W)$ which describes the user utility considering both average service delay and offloading throughput. ϖ is a parameter that accounts for the preference of the VU on the offloading throughput (or cost saved) with respect to the average service delay. Figure 4.5d shows the utility with respect to D_t, with $\varpi = 2.5$. Results of maximum utility and corresponding K are summed in Table 4.1. It can be seen that if the delay constraint is loose (e.g., $D_t = 800$ s), users might care offloading throughput (or cost saved) more than the delay. They may choose larger K for a loose delay constraint to achieve higher offloading throughput and get the maximum utility. On the other hand, if the delay constraint is strict (e.g., $D_t = 400$ s), a smaller K is more likely to be chosen for lower average service delay.

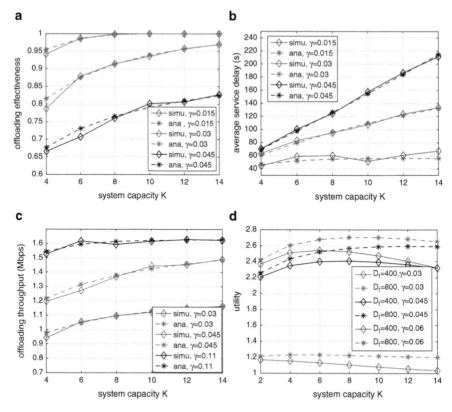

Fig. 4.5 Simulation and theoretical results. (**a**) Offloading effectiveness; (**b**) average service delay; (**c**) offloading throughput; (**d**) user utility. $\varpi = 2.5$

Table 4.1 Maximum utility of VUs

$[K, U_m]$	$\gamma = 0.03$	$\gamma = 0.045$	$\gamma = 0.06$
$D_t = 400\,s$	[2, 1.17]	[8, 2.41]	[6, 2.54]
$D_t = 800\,s$	[4, 1.23]	[12, 2.59]	[8, 2.71]

4.5   Summary

In this chapter, we have theoretically investigated the performance of vehicular WiFi offloading. We have modeled the data service requests of VUs as an M/G/1/K queue, derived the probability distribution of the effective service time and analyzed the performance metrics of vehicular WiFi offloading, i.e., average service delay and offloading effectiveness by the queueing analysis. Simulation results have validated the analysis, and shown the relationship between average service delay and offloading effectiveness.

References

1. KPMG's global automotive executive survey (2012) [Online]. Available: http://www.kpmg.com/GE/en/IssuesAndInsights/ArticlesPublications/Documents/Global-automotive-executive-survey-2012.pdf
2. Seemless wi-fi offload: From vision to reality (2013) [Online]. Available: http://www.enterprisechannelsmea.com/FileManagerImages/12047.pdf
3. Tan WL, Lau WC, Yue O, Hui TH (2011) Analytical models and performance evaluation of drive-thru internet systems. IEEE J Sel Areas Commun 29(1):207–222
4. Luan T, Cai L, Chen J, Shen X, Bai F (2014) Engineering a distributed infrastructure for large-scale cost-effective content dissemination over urban vehicular networks. IEEE Trans Veh Technol 63(3):1419–1435
5. Fiems D, Maertens T, Bruneel H (2008) Queueing systems with different types of server interruptions. Eur J Oper Res 188(3):838–845
6. Abate J, Choudhury GL, Whitt W (2000) An introduction to numerical transform inversion and its application to probability models. In: Computational probability. Springer, Berlin, pp 257–323
7. Bose SK (2001) An introduction to queueing systems. Springer, Berlin
8. Tiger/line shapefiles (2006) [Online]. Available: http://www.census.gov/geo/maps-data/data/tiger-data.html
9. Härri J, Filali F, Bonnet C, Fiore M (2006) VanetMobiSim: generating realistic mobility patterns for VANETs. In: Proceedings of ACM VANET, San Francisco, September 2006

Chapter 5
Conclusions and Future Directions

In this chapter, we summarize the monograph, and provide future research directions.

5.1 Conclusions

In this monograph, we have investigated the utilization of opportunistic spectra for VANETs. Based on the analysis and discussion provided throughout this monograph, we present the following remarks.

- VANETs have faced an increasingly severe problem called spectrum scarcity, which limits the performance of the VANETs and the evolvement of new services and applications, and therefore it is of great importance to provide VANETs with more communication bandwidth resources. Besides the dedicated spectrum for VANETs, more spectrum bands can be utilized for VANETs in an opportunistic manner, such as ISM bands, TV white spaces, cellular bands, etc. How to efficiently utilized these opportunistic spectrum resources is a critical topic for the development of VANETs.
- Cognitive radio technology has been widely studied and proved to be an efficient method to utilize licensed spectrum bands for unlicensed users. Under a urban grid-like street pattern and regular primary transmitter coverage area, the channel availability of an arbitrary moving vehicle can be modeled as exponential distribution. Through the statistics of the channel availability, a game-theoretic spectrum access scheme is proposed for vehicles to access different channels in a distributed manner. The impact of vehicle density and channel diversity on the performance of the proposed channel access scheme has been studied.
- WiFi is envisioned as an efficient and cost-effective method for high-speed wireless access. The WiFi offloading performance for vehicular users is analyzed

© The Author(s) 2016
N. Cheng, X. (Sherman) Shen, *Opportunistic Spectrum Utilization in Vehicular Communication Networks*, SpringerBriefs in Electrical and Computer Engineering,
DOI 10.1007/978-3-319-20445-1_5

using an M/G/1/K queue model. The relation between offloading effectiveness and average service delay is theoretically obtained. The simulation results indicate that generally the delay tolerance of users can improve the offloading effectiveness and therefore save the money.

5.2 Future Research Directions

This monograph presents the preliminary results on the utilization of WiFi for VANETs data delivery, including CR and WiFi, including the study of the characteristics of the opportunistic spectrum in VANETs, such as the channel availability in Chap. 3, and focusing on the performance under some basic utilization strategy, such as the offloading performance and average service delay in Chap. 4. In the future, we intend to investigate exploiting licensed cellular spectrum through D2D communication technology. In addition, we plan to design an integrated communication framework that can effectively incorporate all kinds of available opportunistic spectrums so that VUs can adaptively access different spectrums. Specifically, our future research plan is described as follows.

5.2.1 Exploiting D2D Communication for VANETs

By utilizing the proximity of mobile users and direct data transmission, D2D communications can increase spectral efficiency, offload the cellular network, and reduce communication delay for mobile users. However, there is only limited number of research works that utilize D2D communication in vehicular communication networks. In the future, we plan to employ D2D communication to fulfill some VANETs applications, in order to on one hand offload the cellular network, and to improve the performance of the applications on the other hand.

5.2.1.1 D2D Communications for Content Distribution

Content distribution to VUs is an emerging solution to improve the road safety and driving experience [1]. The types of the content may include advertisement, maps with traffic statistics, road situation report, music and video, etc. Normally, the content is distributed by the cellular network to VUs. However, the large size of some contents, such as videos, may congest the cellular network, if they are requested by a large number of VUs. Therefore, a feasible solution is that the popular contents can be downloaded by only part of vehicles who request, and shared among each other via D2D communications. In this way, cellular spectrum efficiency can be greatly improved, and the cellular network can be offloaded. Another type of content is user generated content that should be shared by neighboring vehicles, such as in a vehicle platoon, the lead vehicle can stream the video recorded by the front camera to platoon members behind, to inform the road condition and avoid accident due to hard break. In traditional cellular network, the content delivery is carried out in a two-hop manner, i.e., vehicle-BS-vehicle, which may lead to extra communication delay. Thus, utilizing D2D communication can not only offload the cellular network, but also reduce the communication delay and improve the performance of the content distribution.

The main issue in exploiting D2D communication for VANETs content distribution is mobility. In D2D communication, the interference should be efficiently managed by resource allocation, mode selection, etc., in order to avoid harmful interference from D2D transmissions to cellular uplink/downlink transmissions, and vise versa. However, the high mobility of vehicles may result in great difficulty in managing the interference. For example, the interference relationships between D2D links and cellular links can be represented by the interference-aware graph, such as in [2]. However, due the vehicle mobility, the interference relationships can be highly dynamic due to the varying positions of vehicles and dynamic channel conditions, which may make it very difficult to formulate a relatively stable interference graph. Thus, to address this issue, we plan to design a dynamic joint resource allocation and mode selection D2D communication scheme that takes the vehicle mobility and QoS requirements of different applications into consideration, in order to efficiently manage interference, increase spectral efficiency, and offload the cellular network.

5.2.1.2 D2D Communications for Safety-Related Applications

Safety applications play a very important role in road safety, such as cooperative collision avoidance and emergency warning message. For safety applications, it is crucial that related messages should be reliably transmitted to receivers who are supposed to obtain the messages within a certain (normally very short) delay. However, in VANETs, it is suggested that the safety-related message dissemination should be carried out over the 10 MHz DSRC control channel without any coordination. Since most safety applications function safety message broadcasting, it is very likely that the safety messages are collided and lost, and a lot of retransmissions are required. This may degrade the reliability of safety applications and introduce extra delay. Thus, with the central control functions of cellular networks, D2D communication can be utilized to improve the performance of safety applications.

We assume that each vehicle is equipped with two radios—DSRC/WiFi radio and cellular radio. Consequently, each vehicle has three communication modes: D2D mode, cellular mode and outband (DSRC) mode. We plan to design a dynamic D2D safety message dissemination scheme that combines resource allocation and mode selection. First, due to the high mobility of vehicles, the resource allocation and mode selection is carried out dynamically to avoid harmful interference. Secondly, the resource allocation and mode selection should consider the specific requirements of safety applications. For example, some safety applications strictly require low communication delay, such as traffic signal violation warning and left turn assistance only allow a delay less than 100 ms. Therefore, the cellular mode should not be considered for such application to guarantee the delay requirement. On the other hand, some safety applications require long transmission range, such as approaching emergency vehicle warning and emergency vehicle signal preemption may require up to 1000 m transmissions. As discussed above, long distance D2D transmission requires larger transmit power, and may result in higher interference. As a result, the cellular mode and outband mode may be preferred.

5.2.2 Opportunistic Communication Framework

An integrated framework that can effectively incorporate all kinds of available opportunistic spectrum is required for VANETs, so that VUs can adaptively access different spectrums. First, due to the movement of vehicles, the availability of the opportunistic transmission spectrums is changing spatially and temporally, which may lead to different availability of different spectrum. Thus, better knowledge of the spectrum availability will lead to better utilization of the opportunistic spectrums. Second, the targets and requirements of accessing different spectrums are varying. For licensed spectrum, the target is to use the vacant bands to improve the communication performance as well as to offload the cellular network, while avoiding or minimizing the interference to primary transmissions. For WiFi access, the spectrum bands are free to use, and the target is to maximize the utility considering cost and user satisfaction (e.g., delay), and to improve the offloading performance given the deployed WiFi network.

One possible solution is that the MNO centrally schedules the opportunistic spectrum usage of VUs on the road. Such a centralized scheduling scheme can be effective based on the following observations.

- The MNO is capable of obtaining the availability of the opportunistic spectrums. For licensed bands, e.g., TV white spaces, the MNO can deploy super WiFi networks using IEEE 802.22 or IEEE 802.11f standards. By doing so, the deployed networks can provide wireless Internet access to vehicles, and offload the congested cellular network. On the other hand, according to the standards, the MNO can access the geolocation database to fetch the permissible frequencies and operating parameters stored by geographic location. For unlicensed bands, the WiFi deployment is known to the MNO if it deploys the WiFi network (i.e., carrier-WiFi network). Moreover, the MNO can also obtain and store the information of WiFi APs deployed by individuals or other corporations and organizations which VUs may not be able or willing to obtain and store.
- Since the MNO can communicate all VUs through the ubiquitous cellular network, it can obtain the QoS requirement of VUs, and use it to scheduling transmissions through different opportunistic spectrums. For example, VUs can bid the acceptable prices in terms of different delays (QoS levels) to the MNO,

and the MNO collects the bids, considers the availability of different spectrums and the vehicle mobility, and then make the scheduling results, which can be seen in [3].

Another possible solution is a hybrid and distributed communication framework which integrates both licensed and unlicensed spectrums. When super WiFi infrastructure is not available, licensed spectrum can be used for V2V communication in cooperative downloading. This is because the V2V links could be easily congested, blocked and faded using the DSRC band, while licensed spectrum such as TVWS can provide improved capacity and better coverage. In the hybrid transmission framework, contents can be downloaded from cellular or WiFi network, and shared among VUs through licensed spectrum. In this way, the capacity of the whole network can be greatly improved, and the congestion in cellular network can be alleviated. In addition, VUs can achieve high utility since the access cost to cellular and WiFi network can be saved by obtaining contents from nearby VUs through cognitive radio.

References

1. Zhang D, Yeo C (2013) Enabling efficient WiFi-based vehicular content distribution. Eur J Oper Res 24(3):479–492
2. Zhang R, Cheng X, Yang L, Jiao B (2013) Interference-aware graph based resource sharing for device-to-device communications underlaying cellular networks. In: Proceedings of IEEE WCNC, Shanghai, April 2013, pp 140–145
3. Zhuo X, Gao W, Cao G, Hua S (2014) An incentive framework for cellular traffic offloading. IEEE Trans Mob Comput 13(3):541–555

Printed in the United States
By Bookmasters